Disclaimer

The publisher of this book is by no way associated with the National Institute of Standards and Technology (NIST). The NIST did not publish this book. It was published by 50 page publications under the public domain license.

50 Page Publications.

Book Title: Evaluation of the Ability of Fire Dynamic Simulator to Simulate Positive Pressure Ventilation in the Laboratory and Practical Scenarios

Book Author: Stephen Kerber

Book Abstract: Positive Pressure Ventilation (PPV) is a tactic that is used on fire grounds across the world everyday, both to improve tenability after the extinguishment of a fire and/or offensively during fire attack to improve firefighting conditions. PPV has proven that it can be a useful tool on the fire ground, but it can also kill or injure fire fighters and civilians if used improperly. Data from three full-scale experiments are compared with simulations completed with the computational fluid dynamic model Fire Dynamic Simulator (FDS). The full-scale experiments characterize a Positive Pressure Ventilation (PPV) fan in an open atmosphere, in a simple room geometry and in a room fire. All experiments qualify and quantify the comparison of the experimental results with the FDS results. A concluding scenario is modeled utilizing the calibration of the full-scale experiments to examine the effects of PPV on a fire in a two-story, colonial style house.

Citation: NIST Interagency/Internal Report (NISTIR) - 7315

Keyword: air flow;FDS;fire dynamic simulator;fire fighter;positive pressure ventilation;PPV;tactics;ventilation

NISTIR 7315

Evaluation of the Ability of Fire Dynamic Simulator to Simulate Positive Pressure Ventilation in the Laboratory and Practical Scenarios

Stephen Kerber

National Institute of Standards and Technology
Technology Administration, U.S. Department of Commerce

NISTIR 7315

Evaluation of the Ability of Fire Dynamic Simulator to Simulate Positive Pressure Ventilation in the Laboratory and Practical Scenarios

Stephen Kerber
Building and Fire Research Laboratory
National Institute of Standards and Technology
Gaithersburg, MD 20899-8661

April 2006

Department of Homeland Security
Michael Chertoff, *Secretary*
Federal Emergency Management Agency
R. David Paulison, *Under Secretary of Emergency Preparedness and Response*
U.S. Fire Administration
Charlie Dickinson, *Deputy Administrator*

U.S. Department of Commerce
Carlos M. Gutierrez, *Secretary*
Technology Administration
Michelle O'Neill, *Acting Under Secretary of Commerce for Technology*
National Institute of Standards and Technology
William A. Jeffrey, *Director*

Evaluation of the Ability of Fire Dynamics Simulator to Simulate Positive Pressure Ventilation in the Laboratory and Practical Scenarios

Stephen Kerber

Abstract

Positive Pressure Ventilation (PPV) is a tactic that is used on fire grounds worldwide everyday, both to improve tenability after the extinguishment of a fire and/or to improve firefighting conditions during fire attack. PPV has proven that it can be a useful tool on the fire ground, but if used improperly it can also kill or injure fire fighters and civilians. Data from three full-scale experiments are compared with simulations completed with the computational fluid dynamic model Fire dynamics simulator (FDS). The full-scale experiments characterize a PPV fan in an open atmosphere, in a simple room geometry and in a room fire.

All experiments qualify and quantify the comparison of the experimental results with the FDS results. A concluding scenario is modeled utilizing the calibration of the full-scale experiments to examine the effects of PPV on a fire in a two-story, colonial style house.

Disclaimer

Certain trade names and company products are mentioned in the text or identified in an illustration in order to specify adequately the experimental procedure and equipment used. In no case does such identification imply recommendation or endorsement by the National Institute of Standards and Technology, nor does it imply that the products are necessarily the best available for the purpose.

Acknowledgements

The author would like to thank staff members of the Building and Fire Research Laboratory at NIST: William D. Walton, Daniel Madrzykowski, Kevin McGrattan, and Glenn Forney for their technical assistance and Robert Vettori, David Stroup, William Twilley, Roy McLane, Andrew Milliken, Jessica Naff, Laurean DeLauter, Ed Hnetkovsky, Jack Lee, Robert Anleitner, Jay McElroy and Alexander Maranghides for the support they provided during this set of experiments. In addition, the author expresses gratitude to the United States Fire Administration, part of the Department of Homeland Security, Robert McCarthy, Meredith, Lawler and William Troup for their support of this important research program.

A special thank you to Dr. James Milke for guiding me through this process, and to Dr. Frederick Mowrer and Dr. Arnaud Trouve for participating as part of the advisory committee.

Table of Contents

Acknowledgements..ii
Table of Contents..iii
List of Tables..v
List of Figures...vi
Chapter 1: Introduction..1
   1.1   Positive Pressure Ventilation..1
   1.2   Fire Dynamics Simulator...2
   1.3 Smokeview..3
Chapter 2: Mapping the PPV Velocity Flow Field...4
   2.1 Experimental   Description..4
      2.1.1 Experimental   Facility..4
      2.1.2 Experiment   Components...4
      2.1.3 Experiment   Procedure...7
      2.1.4   Flow Visualization Experiment..8
   2.2 Computer   Simulation...9
      2.2.1 Domain..9
      2.2.2   Geometry and Vents...9
      2.2.3 Output   Files...10
   2.3 Results...11
Chapter 3: Simple Room Experiment..16
   3.1 Experimental   Description..16
      3.1.1 Experimental   Facility..16
      3.1.2 Experiment   Components...16
      3.1.3 Experiment   Layout..17
      3.1.4 Experiment   Procedure...18
      3.1.5   Flow Visualization Experiment..19
   3.2 Computer   Simulation..20
      3.2.1 Domain..20
      3.2.2   Geometry and Vents...22
      3.2.3 Output   Files...22
   3.3 Results...22
   3.4   Mapping and Simple Room Summary..26
Chapter 4: Room Fire Experiments..30
   4.1 Experimental   Description..30
      4.1.1 Instrumentation...33
      4.1.2 Fuel   Load..37
      4.1.3 Procedure...44
   4.2 Experimental   Results..46
      4.2.1   Heat Release Rate..50
      4.2.2   Room Gas Temperature..53
      4.2.3   Doorway Gas Temperature..55
      4.2.4   Window Gas Temperature..57

  4.2.5 Corridor Gas Temperature..59
  4.2.6 Room Differential Pressure..61
  4.2.7 Window Gas Velocity...63
  4.2.8 Doorway Gas Velocity..65
 4.3 Computer Simulation..67
  4.3.1 Domain..67
  4.3.2 Geometry...70
  4.3.3 Materials..70
  4.3.4 Vents and Ignition Source..71
  4.3.5 Output Files..72
 4.4 Simulation Results..72
  4.4.1 Naturally Ventilated Simulation..72
  4.4.2 Positive Pressure Ventilated Simulation...85
 4.5 Discussion...96
 4.6 Room Fire Summary..98

Chapter 5: Colonial House Practical Scenario...100
 5.1 Scenario Overview...100
 5.2 Computer Simulations...100
  5.2.1 Domain..100
  5.2.2 Geometry...103
  5.2.3 Vents..107
  5.2.4 Materials..108
  5.2.5 Fire Source...110
  5.2.6 Output Files..110
 5.3 Results...111
  5.3.1 Fire Growth and Smoke Spread..111
  5.3.2 Heat Release Rate..114
  5.3.3 Temperature..115
  5.3.4 Oxygen..119
  5.3.5 Velocity...123
 5.4 Colonial House Summary..128

Chapter 6: Uncertainty...130
 6.1 Experimental..130
 6.2 Fire Dynamics Simulator...133

Chapter 7: Conclusion..134
References..136

List of Tables

Table 1. Comparison of Experimental and FDS Velocity Mapping Results.............11
Table 2. Comparison of Experimental and FDS Simple Room Results....................23
Table 3. Summary of FDS Characteristics Needed to Model the Fan Flow..............26
Table 4. Summary of Multi-block Characteristics Needed to Model Simple Room Flow...26
Table 5. Fire Load Weights..38
Table 6. Experimental Procedure..44
Table 7. Observations..47
Table 8. Room Fire FDS Input Material Properties..71
Table 9. Colonial FDS Input Material Properties...109
Table 10. Mapping and Simple Room Experimental Uncertainty..........................131
Table 11. Room Fire Experimental Uncertainty...132

List of Figures

Figure 1-1. Two Common PPV Fans...2
Figure 1-2. PPV Cone of Air [4]..2
Figure 2-1. The Flow Characterization Grid...5
Figure 2-2. Flow Visualization Threads..5
Figure 2-3. Front and Back of PPV Fan..6
Figure 2-4. Anemometer..6
Figure 2-5. Anemometer in Indexer on Track...6
Figure 2-6. Velocity Mapping Experimental Layout...7
Figure 2-7. Velocity Mapping Measurement Points..8
Figure 2-8. Smoke Generator Flowing...8
Figure 2-9. Layout for Supplemental Experiment..8
Figure 2-10. Visualization of PPV Flow..9
Figure 2-11. Fan and Smoke Generator...9
Figure 2-12. FDS PPV Fan Visualized in Smokeview.......................................10
Figure 2-13. FDS Layout Visualized with Smokeview......................................11
Figure 2-14. Fine Grid Cell Visualization..11
Figure 2-15. Smokeview Visualization of FDS PPV Flow Pattern....................12
Figure 2-16. Experimental Velocities 1.8 m (6 ft) From the Fan.......................13
Figure 2-17. FDS Velocities 1.8 m (6 ft) From the Fan.....................................13
Figure 2-18. Experimental Velocities 2.4 m (8 ft) From the Fan.......................14
Figure 2-19. FDS Velocities 2.4 m (8 ft) From the Fan.....................................14
Figure 2-20. Experimental Velocities 3.1 m (10 ft) From the Fan.....................15
Figure 2-21. FDS Velocities 3.1 m (10 ft) From the Fan...................................15
Figure 3-1. Floor Plan of Simple Room..17
Figure 3-2. Simple Room Experimental Layout...17
Figure 3-3. Experimental Layout Looking at Room Inlet and Outlet.................17
Figure 3-4. Inlet and Outlet From a Different Perspective.................................18
Figure 3-5. Doorway (Inlet) Measurements Points..19
Figure 3-6. Window (Outlet) Measurement Points..19
Figure 3-7. Visualization of Supplemental Experiment as Soon as Room is
    Pressurized...20
Figure 3-8. Visualization of Supplemental Experiment Once Constant Flow is
    Achieved...20
Figure 3-9. FDS Layout for Simple Room Experiment......................................21
Figure 3-10. Grid Cell Visualization, Multi-blocking..21
Figure 3-11. FDS Layout Looking at Room Inlet and Outlet.............................22
Figure 3-12. FDS Layout with Fan Operating..23
Figure 3-13. Experimental Velocity Measurements in the Doorway (Inlet)......24
Figure 3-14. FDS Velocities in the Doorway (Inlet)..24
Figure 3-15. Experimental Measurements in the Window (Outlet)...................25
Figure 3-16. FDS Velocities in the Window (Outlet)..25
Figure 3-17. Smokeview Visualization of FDS PPV Flow Pattern....................27
Figure 3-18. Experimental Visualization of the PPV Flow Pattern...................27

Figure 3-19. FDS Visualization of Supplemental Experiment as Soon as Room is Pressurized..28
Figure 3-20. Visualization of Supplemental Experiment as Soon as Room is Pressurized..28
Figure 3-21. FDS Visualization of Supplemental Experiment Once Constant Flow is Achieved..29
Figure 3-22. Visualization of Supplemental Experiment Once Constant Flow is Achieved..29
Figure 4-1. Experimental Floor Plan..31
Figure 4-2. External View of Window in Closed Position..32
Figure 4-3. External View of Open Corridor Doorway...32
Figure 4-4. Furniture Floor Plan...33
Figure 4-5. Instrumentation Floor Plan (TC: thermocouples, BDP: bidirectional probes)...35
Figure 4-6. Doorway and Window Probe and Thermocouple Combination............36
Figure 4-7. View of Doorway From Inside the Room into Corridor and Window with Instrumentation..36
Figure 4-8. Bidirectional Probe and Thermocouple Combination............................37
Figure 4-9. Water Cooled Camera...37
Figure 4-10 - Book Case...39
Figure 4-11 - Book Case Dimensions...39
Figure 4-12 - Desk and Monitor...40
Figure 4-13 - Desk Dimensions..40
Figure 4-14 - Monitor...41
Figure 4-15 - Monitor Dimensions...41
Figure 4-16 - Chair...42
Figure 4-17 - Chair Dimensions...42
Figure 4-18 – Bunk Bed...43
Figure 4-19 – Bunk Bed Dimensions...43
Figure 4-20. Ignition Setup..45
Figure 4-21. Electrically Activated Matchbook Locations..45
Figure 4-22. Exterior View of Doorway to Corridor After the Start of Forced Ventilation (470 s)..48
Figure 4-23. Exterior View of Window After the Start of Forced Ventilation (380 s)...48
Figure 4-24. Doorway During Natural Ventilation Experiment (645 s)...................49
Figure 4-25. Window During Natural Ventilation Experiment (470 s)....................49
Figure 4-26. Heat Release Rate...51
Figure 4-27. Heat Release Rate Detail For 200 s Following Peak Output...............52
Figure 4-28. Total Heat Released..52
Figure 4-29. PPV Room Temperatures, Distances Measured From Ceiling.............54
Figure 4-30. Natural Ventilation Room Temperatures...55
Figure 4-31. PPV Doorway Temperatures..56
Figure 4-32. Natural Ventilation Doorway Temperatures.......................................57
Figure 4-33. PPV Window Temperatures...58
Figure 4-34. Natural Ventilation Window Temperatures..59

Figure 4-35. PPV Corridor Temperatures..60
Figure 4-36. Natural Ventilation Corridor Temperatures..61
Figure 4-37. PPV Room Differential Pressure..62
Figure 4-38. Natural Ventilation Room Differential Pressure..................................63
Figure 4-39. PPV Window Velocities...64
Figure 4-40. Natural Ventilation Window Velocities...65
Figure 4-41. PPV Room Doorway Velocities...66
Figure 4-42. Natural Ventilation Room Doorway Velocities...................................67
Figure 4-43. FDS Naturally Ventilated Domain...68
Figure 4-44. FDS PPV Ventilated Domain...69
Figure 4-45. Grid Cell Visualization...69
Figure 4-46. Rectangular Geometries...70
Figure 4-47. Location of Ignition Source...72
Figure 4-48. Bunkbed View at Time of Ignition (0 s)..73
Figure 4-49. Doorway and Window View at Time of Ignition (0 s).......................73
Figure 4-50. Fire Starting on Corner of Mattress (90 s)...74
Figure 4-51. Flames Involving Bunkbed (230 s)..74
Figure 4-52. Onset of Flashover (275 s)...75
Figure 4-53. Visibility Lost in Bunk Bed View (300 s)...75
Figure 4-54. Combustion Products Flow From Corridor Doorway (330 s).............76
Figure 4-55. Flames Extend From Corridor Doorway Once Window is Opened
  (360 s)...76
Figure 4-56. Flames From Window (360 s)...77
Figure 4-57. Flames Continue From Window (430 s)...77
Figure 4-58. Fire in Decay Stage (540 s)..78
Figure 4-59. Room Continues to Burn (720 s)...78
Figure 4-60. FDS and Experimental Naturally Ventilated Heat Release Rate.........79
Figure 4-61. FDS and Experimental Naturally Ventilated Total Heat Released.....80
Figure 4-62. FDS Naturally Ventilated Room Temperatures...................................81
Figure 4-63. FDS Naturally Ventilated Room Doorway Temperatures...................82
Figure 4-64. FDS Naturally Ventilated Corridor Doorway Temperatures...............83
Figure 4-65. FDS Naturally Ventilated Window Velocities....................................84
Figure 4-66. FDS Naturally Ventilated Room Doorway Velocities........................85
Figure 4-67. External Door View With Fan Prior to Ignition (0 s)..........................86
Figure 4-68. Bunkbed View as Flames Involve Corner of Mattress (60 s).............86
Figure 4-69. Doorway and Window View as Flames Spread to Top Mattress
  (170 s)...86
Figure 4-70. Flames Involving Bunk Bed (240 s)..87
Figure 4-71. Visibility Lost in Bunk Bed View (300 s)...87
Figure 4-72. Thick Smoke Flows From Corridor Doorway (300 s).........................88
Figure 4-73. Doorway and Window View, From Inside the Room, Obstructed
  by Flames (360 s)...88
Figure 4-74. Doorway View 10 s After Fan is Turned On (360 s)..........................88
Figure 4-75. Flames From Window (360 s)...89
Figure 4-76. Flames Continue From Window (400 s)...89
Figure 4-77. Combustion Products Forced into Room by Fan (410 s)....................90

Figure 4-78. Fire in Decay Stage (445 s)..90
Figure 4-79. Fan Forcing Flow Through Room (500 s)...90
Figure 4-80. FDS and Experimental Positive Pressure Ventilated Heat Release Rate..92
Figure 4-81. FDS Heat Release Rate Comparison of natural and PPV ventilation...92
Figure 4-82. FDS PPV Ventilated Room Temperatures..93
Figure 4-83. FDS PPV Ventilated Room Doorway Temperatures.............................94
Figure 4-84. FDS PPV Ventilated Corridor Doorway Temperatures........................95
Figure 4-85. FDS PPV Ventilated Window Velocities..95
Figure 4-86. FDS PPV Ventilated Room Doorway Velocities...................................96
Figure 5-1. Colonial House and Grid Locations...101
Figure 5-2. Display of Grid Cell Size...102
Figure 5-3. Colonial House and PPV Fan Placement..102
Figure 5-4. Floor Plan of First Floor..103
Figure 5-5. Furniture Locations on First Floor..104
Figure 5-6. Smokeview Display of First Floor..104
Figure 5-7. Floor Plan of Second Floor..105
Figure 5-8. Furniture Locations on Second Floor...105
Figure 5-9. Smokeview Display of Second Floor...106
Figure 5-10. Basement Floor Plan..106
Figure 5-11. Smokeview Display of Basement...107
Figure 5-12. Front and Rear View of House..108
Figure 5-13. Location of Room Vents with Roof Removed.....................................108
Figure 5-14. View of Interior from Inside the Front Door..109
Figure 5-15. Fire Source in Bedroom...110
Figure 5-16. Growth of Fire Prior to Ventilation, 60 s to 240 s................................112
Figure 5-17. Comparison of Simulations at 250 s (Natural left, PPV right)............113
Figure 5-18. Comparison of Simulations at 360 s (Natural left, PPV right)............113
Figure 5-19. Comparison of Simulations at 360 s (Natural left, PPV right)............113
Figure 5-20. Comparison of Simulations at 600 s..114
Figure 5-21. Colonial House Heat Release Rate Comparison..................................115
Figure 5-22. Comparison of Second Floor Temperatures 100 s to 300 s, Naturally Ventilated (Left) and PPV Ventilated (Right)...................................117
Figure 5-23. Comparison of Second Floor Temperatures 400 s to 600 s, Naturally Ventilated (Left) and PPV Ventilated (Right)...................................118
Figure 5-24. Comparison of First Floor Temperatures 200 s to 600 s, Naturally Ventilated (Left) and PPV Ventilated (Right)...................................119
Figure 5-25. Comparison of Second Floor Oxygen Volume Fractions 100 s to 300 s, Naturally Ventilated (Left) and PPV Ventilated (Right).......................121
Figure 5-26. Comparison of Second Floor Oxygen Volume Fractions 300 s to 600 s, Naturally Ventilated (Left) and PPV Ventilated (Right).......................122
Figure 5-27. Comparison of First Floor Oxygen Volume Fractions 200 s to 600 s, Naturally Ventilated (Left) and PPV Ventilated (Right).......................123
Figure 5-28. Vertical Velocity Slice Through Center of Fan....................................124
Figure 5-29. Velocity Slice Through Center of Vent Window During PPV Ventilation Scenario at 320 s...125

Figure 5-30. Velocity Slice Through Center of Vent Window During Natural Ventilation Scenario at 320 s..125

Figure 5-31. Horizontal Velocity Slice Through Second Floor During PPV Ventilation Scenario at 320 s..126

Figure 5-32. Horizontal Velocity Slice Through Second Floor During Natural Ventilation Scenario at 320 s..126

Figure 5-33. Horizontal Velocity Slice Through First Floor During PPV Ventilation Scenario at 320 s..127

Figure 5-34. Horizontal Velocity Slice Through First Floor During Natural Ventilation Scenario at 320 s..127

Chapter 1: Introduction

Positive Pressure Ventilation (PPV) is a common tactic used by fire departments to ventilate a structure after a fire has been extinguished. Typically this allowed fire fighters to complete salvage and overhaul operations in a less hazardous environment. PPV is also used during a fire attack. The fan is started in coordination with a ventilation opening during the initial phase of the fire attack. This tactic was designed to increase visibility and force heat away from the attack team as they locate and extinguish the fire. PPV has been implemented with some success, but also with some difficulty. It has not been carefully characterized in terms of gas temperatures, gas velocities and mass burning rates. The uncertainties associated with PPV have given rise to several issues: When should PPV be used, and just as important, when should it not be used? What is the best location for the fan and where should the exhaust or vent opening be made? Does PPV provide oxygen to the fire and allow for quicker fire growth? What is the consequence if fire fighters or building occupants are between the fire and the exhaust opening? Are there certain types of construction when PPV should not be used?

As early as 1989, fire department training publications questioned whether PPV fans would intensify a fire by introducing additional oxygen. Carlson [1] responded to this question, stating that it was a possibility, but there was no evidence or research to substantiate his answer. As part of their "Roundtable" discussion in 1999, *Fire Engineering* polled fire chiefs from around the country to determine the extent to which their departments used PPV. Many of the polled departments used PPV, but some did not use it offensively with the explanation that intensifying the fire was a concern and this phenomenon was not well understood [2].

In 2001, Yates conducted an investigation and survey of the usage of PPV in the Tyne and Wear Metropolitan Fire Brigade in the United Kingdom, Salt Lake City, Utah Fire Department in the United States, and Aachen Fire Department in Germany [3]. The survey suggested a few reasons for the lack of implementation of PPV. Two of these reasons were the potential for increased damage to structures and insufficient research data and evidence available to support its benefits. This is further evidence that PPV lacks the firm scientific foundation necessary for salvage, overhaul, and fire suppression operations.

1.1 Positive Pressure Ventilation

PPV is a ventilation technique used by the fire service to remove smoke, heat and other combustion products from a structure. This allows the fire service to perform other tasks in a more tenable environment. PPV fans are commonly powered with an electric or gasoline engine and range in diameter from 0.30 m to 0.91 m (12 in to 36 in) (Figure 1-1). Typically, a PPV fan is placed about 1.8 m to 3.0 m (6 ft to 10 ft) outside the doorway of the structure. It is positioned so that the "cone of air" produced by the fan extends beyond the boundaries of the opening (Figure 1-2). With the doorway within the cone of air, pressure inside the structure increases. An exhaust opening in the structure, such as an opening in the roof or an open window, allows the air to escape due to the difference between the inside and outside air pressure. The smoke, heat and other combustion products are pushed out of the structure and replaced with ambient air [4].

Figure 1-1. Two Common PPV Fans

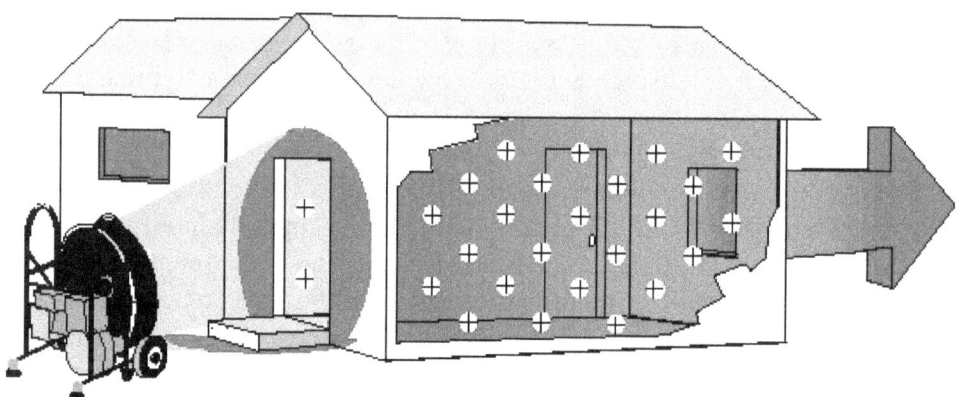

Figure 1-2. PPV Cone of Air [4]

1.2 Fire Dynamics Simulator

The National Institute of Standards and Technology's Fire Dynamics Simulator (FDS) is a computational fluid dynamics (CFD) model of fire-driven fluid flow. It numerically solves a form of the Navier-Stokes equations appropriate for low-speed, thermally driven flow with an emphasis on smoke and heat transport from fires [5]. Version 1 was publicly released in February 2000. The predictions performed here were made with the public pre-release version 4 of the model [6]. Version 4 includes several new features such as multi-blocking which were critical in performing the room fire simulations.

A CFD model requires that the room or building of interest be divided into small three-dimensional rectangular control volumes or computational cells. The CFD model computes the density, velocity, temperature, pressure and species concentration of the gas in each cell as the model steps through time. Based on the laws of conservation of mass, momentum, species and energy, the model tracks the generation and movement of fire gases. Radiative heat transfer is included in the model via the solution of the radiation transport equation for a non-scattering gray gas. All solid surfaces are assigned thermal boundary conditions, in addition to information about the burning behavior of the material. Heat and mass transfer to and from solid surfaces is usually handled with empirical correlations. FDS utilizes material properties of the furnishings, walls, floors, and ceilings to compute fire growth and spread. A complete description of the FDS model is given in references [5, 6].

Inputs required by FDS include the geometry of the structure and furnishings, the computational cell size, the location of the ignition source, the energy release rate of the ignition source, thermal properties of walls, ceilings, floors, furnishings, and the size, location, and timing of door and window openings to the outside. Each of these inputs can significantly influence fire growth and spread.

1.3 Smokeview

Smokeview is a scientific visualization program that was developed to display the results of an FDS model computation [7]. Smokeview allows the viewing of FDS results in three-dimensional snapshots or animations. Smokeview can display contours of temperature, velocity and gas concentration in planar slices. It can also display properties with iso-surfaces that are three-dimensional versions of a constant value of the property. Iso-surfaces are most commonly used to provide a three-dimensional approximation of the flame surface where fuel and oxygen mix and burn.

Chapter 2: Mapping the PPV Velocity Flow Field

This research effort used a series of full-scale experiments to examine how PPV may impact structural ventilation. These same experiments were simulated with FDS to provide more insight into the impact of ventilation on fire behavior. The computer simulations were compared with the full-scale test results. This enabled the validation of the computer simulation and as necessary, identified areas that need improvement. Ultimately, PPV computer simulations could be used to improve fire fighter safety by enabling improved understanding of structural ventilation techniques.

PPV fans are engineered to maximize airflow while allowing the fan to remain light and durable for fire service use. Simply, a blade or impeller pulls air through a shroud creating the airflow. This blade or impeller is driven by an electric or gasoline engine all of which are mounted in a frame. The velocity field that is created is complex due to the speed at which the blade or impeller rotates to achieve a conical flow. These mapping experiments examine FDS's ability to accurately characterize this complex flow.

2.1 Experimental Description

The initial series of experiments served to determine the flow field created by PPV fans. In order to implement a PPV fan in FDS it was necessary to characterize the fan in terms of FDS input parameters such as obstruction and vent locations and orientations. This basic environment was an open atmosphere with the fan located on a stand that had negligible effects on the flow, essentially placing the fan in open space. This placement minimized the impact from obstructions such as walls and doorways on the flow field.

2.1.1 Experimental Facility

These experiments were conducted at NIST's Building and Fire Research Laboratory Large Fire Facility. Each experiment was located in an area inside the facility so that the airflow created by the PPV fan was not affected by external factors such as wind or weather. The facility has the interior dimensions, 36.6 m (120 ft) long, 18.3 m (60 ft) wide and 7.6 m (25 ft) high.

2.1.2 Experiment Components

Grid

A grid frame, 2.44 m x 2.44 m (8 ft x 8 ft), was constructed with 51 mm x 102 mm (2 in x 4 in) wood members to form a square. The corners were reinforced with plywood triangles. Cotton strings approximately 1.6 mm (0.0625 in) in diameter were placed on the frame both vertically and horizontally to form a grid. The center was highlighted by using orange string, while white strings were placed every 0.1m (4 in) for the first 0.31m (12 in) then every 0.15 m (6 in) towards the edge of the frame in both directions. The frame was notched to keep the strings in place (Figure 2-1). Once the strings were all in place, 0.2 m (8 in) long threads were added to each point on the grid. These threads made it possible to view the direction of airflow (Figure 2-2). Finally, black felt was added to the frame, parallel to the threads in order to give a contrasting background to the threads and aid in photographic documentation.

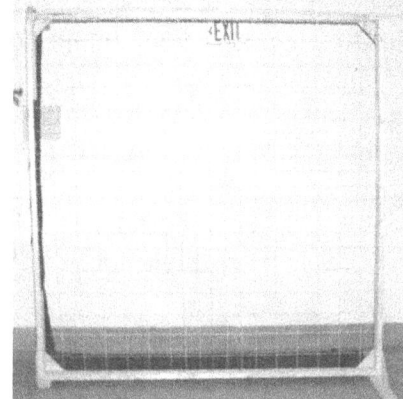

Figure 2-1. The Flow Characterization Grid

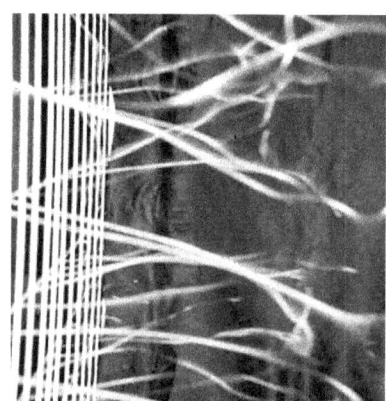

Figure 2-2. Flow Visualization Threads

Fan

The fan used was an 18-inch, variable speed, electric, positive pressure ventilator. The fan had a depth of 0.48 m (18.75 in), width of 0.62 m (24.5 in) and height of 0.62 m (24.5 in). It had a maximum speed of 2200 rpm, a horsepower rating of 746 W (1 hp) and a flow rating of 6.64 m^3/s (14,060 ft^3/min). The fan was mounted on a stand to set the fan at the proper height of 1.28 m (50.5 in), measured from the center of the fan to the ground (Figure 2-3). The grid was positioned so that the center of the fan impeller was aligned with the centerline of the grid.

Figure 2-3. Front and Back of PPV Fan

Anemometer

The digital anemometer used was a 25.4 mm (1 in) diameter vane-type probe (Figure 2-4). It is microprocessor-based with a range of 0.3 m/s (1.0 ft/s) to 35.0 m/s (115 ft/s) with an accuracy of 0.5% of the readings [8].

In order to position the anemometer at various locations an anemometer indexer was fabricated. The indexer allowed the anemometer to be placed at any point on the grid by moving it horizontally and vertically. The indexer also permitted the anemometer to rotate around both the x and y-axis. This rotation allowed the measurement of velocities that were not parallel to the ground and perpendicular to the fan. The indexer moved horizontally across the grid on wheels that were grooved to ride an angle iron track. This track was positioned on the ground parallel with the grid (Figure 2-5).

Figure 2-4. Anemometer   Figure 2-5. Anemometer in Indexer on Track

2.1.3 Experiment Procedure

The grid and track were placed perpendicular to the flow. The fan was positioned so that the flow was centered on the grid 1.28 m (4.2 ft) above the floor. The anemometer indexer was moved along the track and the height of the anemometer adjusted to correspond with the measurement position on the grid (Figure 2-6). The vane of the anemometer was maintained perpendicular to the flow. The fan was started and run for two minutes at the maximum speed of 2200 rpm. The anemometer output was recorded every two seconds for four minutes using 16 second averaging. This process was repeated for the selected points on the grid shown in Figure 2-7.

Figure 2-6. Velocity Mapping Experimental Layout

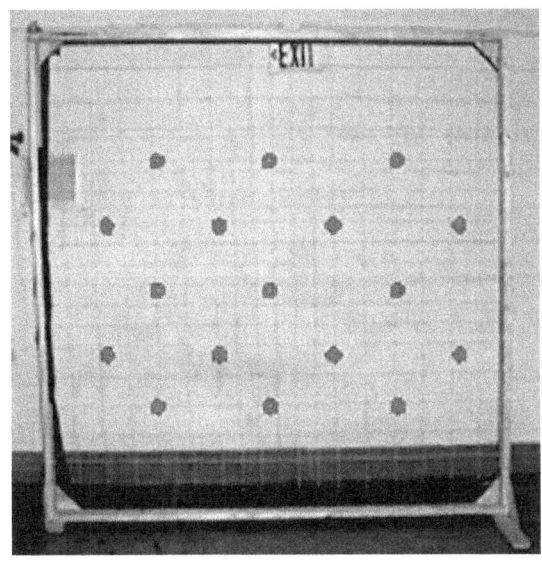

Figure 2-7. Velocity Mapping Measurement Points

2.1.4 Flow Visualization Experiment

In order to visualize and qualify the flow from the fan, a supplemental test was performed. A smoke generator was placed 0.3 m (1 ft) to the rear of the fan (Figure 2-11). The fan was turned on to a speed of 2200 rpm and allowed to reach steady state operating conditions. The smoke generator was then turned on to maximum output and pictures were taken against a black background with heights labeled in 0.3 m (1 ft) intervals and widths labeled in 0.6 m (2 ft) intervals. See Figures 2-8 through 2-11.

Figure 2-8. Smoke Generator Flowing   Figure 2-9. Layout for Supplemental Experiment

Figure 2-10. Visualization of PPV Flow   Figure 2-11. Fan and Smoke Generator

2.2 Computer Simulation

Inputs required by FDS in order to model PPV fans included the following: domain size, computational cell size, vent velocity, vent geometry, fan geometry, slice location and velocity measurement points. Many of the inputs that were used for preliminary FDS runs are explained in the following sections.

2.2.1 Domain

Many issues need to be considered when modeling a PPV fan using FDS. The most important is the computational cell size. It was found that the cells needed to be no larger than 16.4 cm^3 (1 in^3). For cells greater than 18.0 cm^3 (1.1 in^3) the flow from the fan did not form a cone. The cell size in the computation in Figures 2-12 through 2-14 are 16.4 cm^3 (1 in^3).

The next consideration was the domain size. At least 1 m (3.3 ft) to the rear of the fan was needed in order for it to create a cone of air. If the fan were located at or within 1 m (3.3 ft) of the boundary, whether the boundary was open or closed, the fan flow was not predicted accurately.

2.2.2 Geometry and Vents

The obstructions and vents that make up the fan itself needed to be prescribed correctly. FDS only allows rectangular obstructions and vents to be created. This leads to an issue due to the cylindrical nature of the fan shroud. In order to get the proper flow pattern the shroud must be prescribed as a series of obstructions oriented in a circle such as Figure 2-12. The degree of roundness depends of the accuracy that is desired. The shroud in Figure 2-12 yielded results within 15% of the experimental but the shroud can be created more cylindrical if desired.

The velocity planes or vents were prescribed as the interior dimensions of the shroud. Vents were also located on the front of the shroud, opposite the motor and handle. This allowed air to be pulled through the shroud creating a more realistic flow pattern. If the vents are placed to the rear or middle of the shroud the flow pattern would appear linear and unrealistic. The velocity into the vents was also prescribed equal to the maximum speed of the experimental PPV fan, 17.89 m/s (40 mi/h). This input was based on the maximum speed of the experimental PPV fan from the previous section. The prescribed velocity corresponded to a pressure drop in the calculation which created the conical flow pattern. Altering this input allowed the user to characterize the fan operating at different speeds.

The final items to be described were the obstructions in the center of the fan, simulating the center of the blade connected to the shaft, and the motor and handle to the rear of the shroud. Adding these obstructions created a more realistic flow pattern. They affected the air moving through the shroud in the FDS simulation similarly as they did in the experiment. Pulling air past the motor affected the flow pattern significantly so it had to be included in the model.

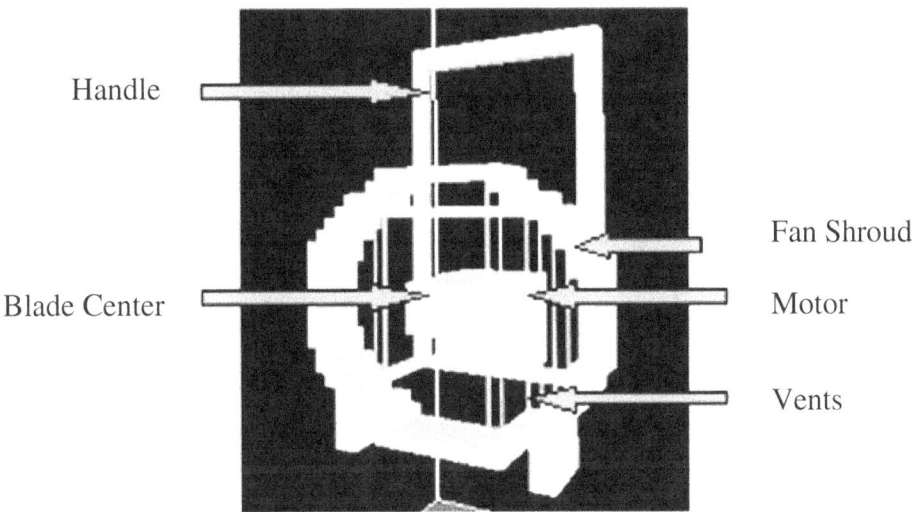

Figure 2-12. FDS PPV Fan Visualized in Smokeview

2.2.3 Output Files

Slice files and prediction points were prescribed. The slice files were placed at the center of the fan in both the horizontal and vertical directions to visualize the flow pattern. Velocity measurement points were placed in the model in conjunction with the measurement points used experimentally, in order to make a comparison of the two at a distance of 1.83 m, 2.44 m, and 3.05 m (6 ft, 8 ft and 10 ft) from the PPV fan.

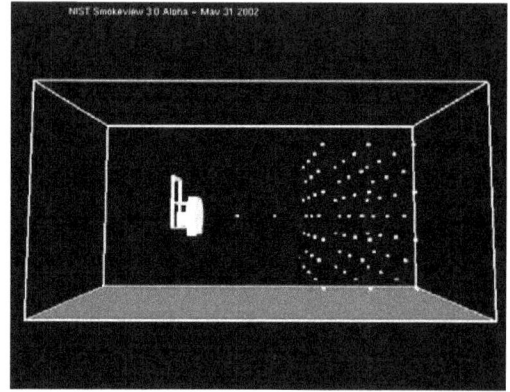
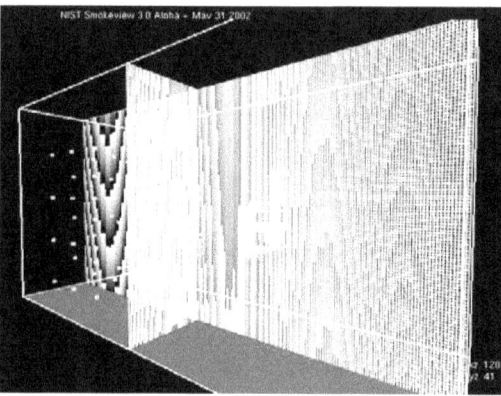

Figure 2-13. FDS Layout Visualized with Smokeview

Figure 2-14. Fine Grid Cell Visualization

2.3 Results

Velocity measurements were taken at the locations depicted in Figure 2-7. These measurements were recorded with the fan at 1.83 m, 2.44 m, and 3.05 m (6 ft, 8 ft and 10 ft) from the anemometer, corresponding to typical distances of a PPV fan from a structure. The magnitudes of the velocities are shown in Figures 2-16, 2-18 and 2-20. The average velocities of the three distances were 2.42 m/s, 2.72 m/s and 3.35 m/s (7.9 ft/s, 8.9 ft/s and 11.0 ft/s), respectively. Many simulations were performed examining the possibilities of creating a PPV fan that closely portrays the actual fan that was used for experimentation. The final fan geometry, Figure 2-15, yielded velocity measurements that are graphed in Figures 2-17, 2-19 and 2-21. The average velocities of the three distances were 2.65 m/s, 3.19 m/s and 3.25 m/s (8.7 ft/s, 10.5 ft/s and 10.7 ft/s) respectively. Average velocities were compared due to the fluctuations in flow across the measurement plane of interest. Velocities were averaged over the duration of the simulation and across the plane of interest. This comparison gave an average difference of slightly less than 10 % (Table 1). The quality of the flow was compared in Figures 2-17 and 2-18.

Table 1. Comparison of Experimental and FDS Velocity Mapping Results

	Average Velocities, m/s (ft/s)		
Distance from fan	1.8 m (6 ft.)	2.4 m (8 ft.)	3.1 m (10 ft.)
Experimental	2.42 (7.9)	2.72 (8.9)	3.35 (11.0)
Fire dynamics simulator	2.65 (8.7)	3.19 (10.5)	3.25 (10.7)
Relative Difference (%) (Exp-FDS)/Exp x 100	8.7	14.7	-3.1

Note: Experimental velocity total expanded uncertainty is ± 14%, See Table 10.

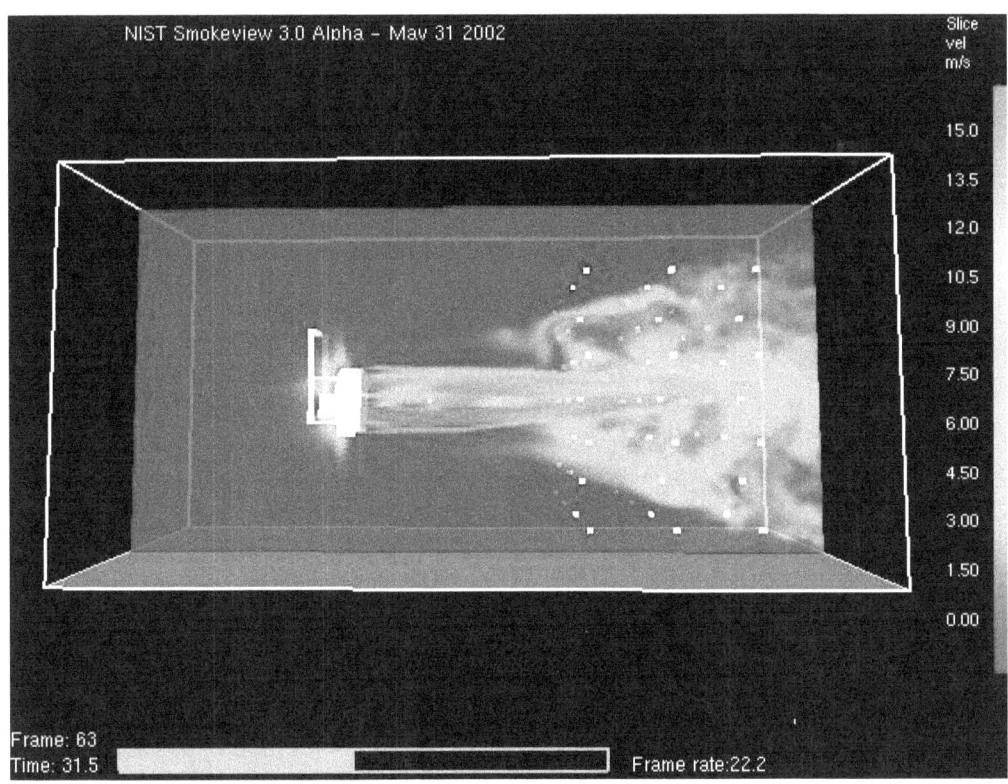

Figure 2-15. Smokeview Visualization of FDS PPV Flow Pattern

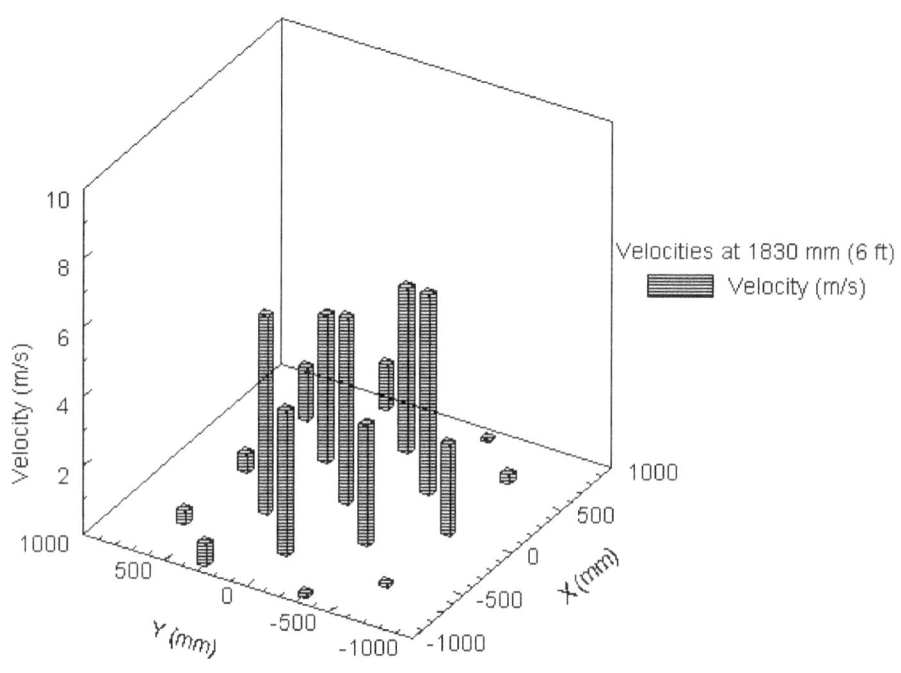

Figure 2-16. Experimental Velocities 1.8 m (6 ft) From the Fan

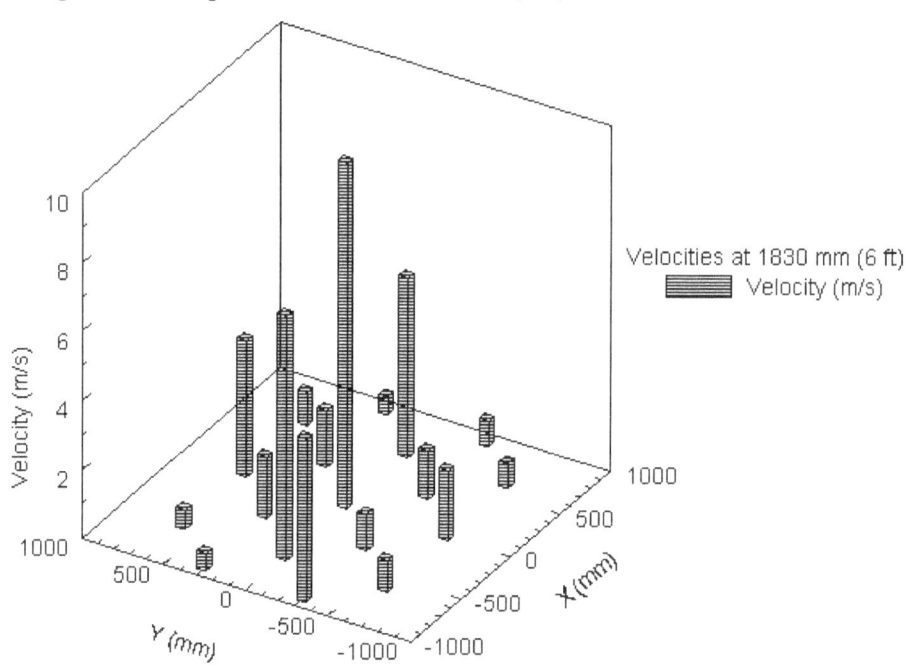

Figure 2-17. FDS Velocities 1.8 m (6 ft) From the Fan

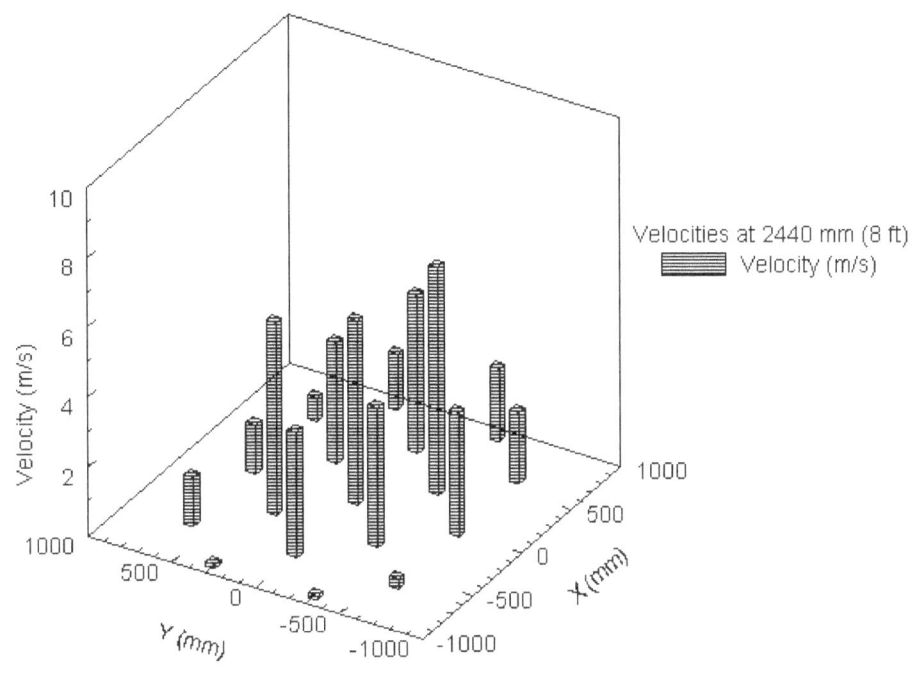

Figure 2-18. Experimental Velocities 2.4 m (8 ft) From the Fan

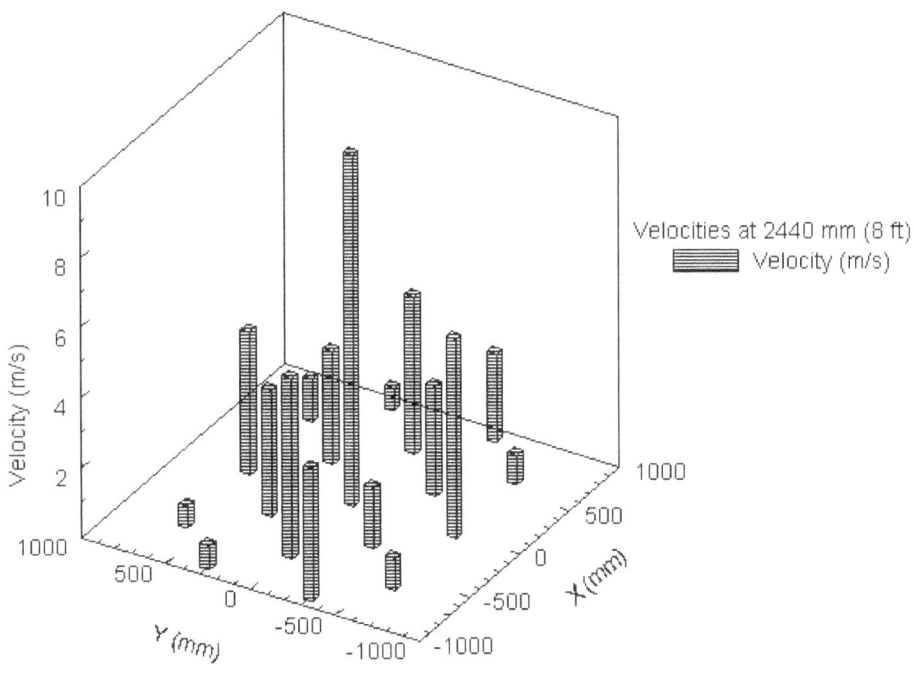

Figure 2-19. FDS Velocities 2.4 m (8 ft) From the Fan

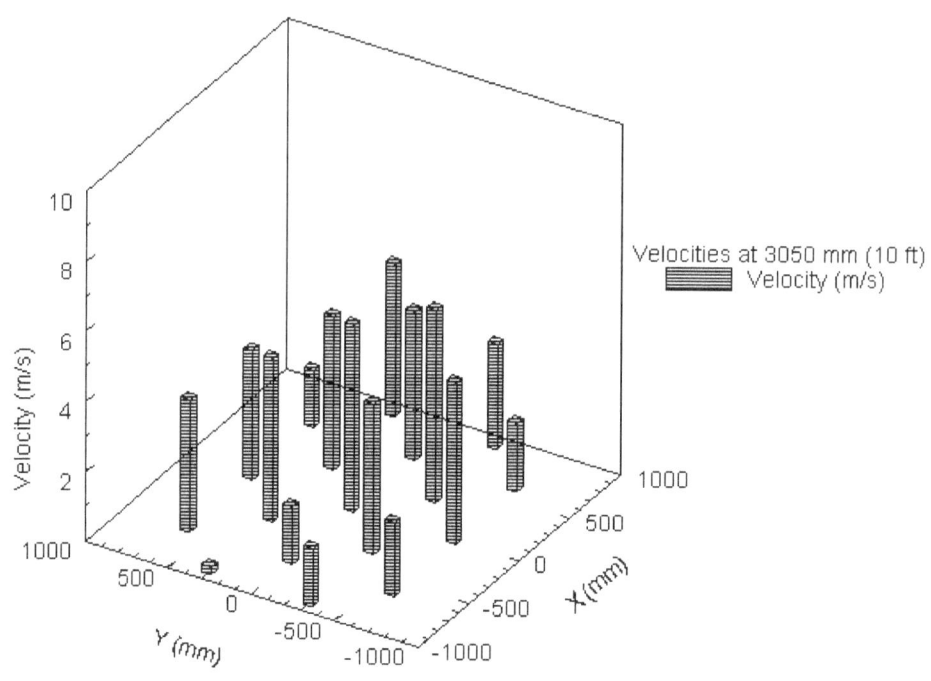

Figure 2-20. Experimental Velocities 3.1 m (10 ft) From the Fan

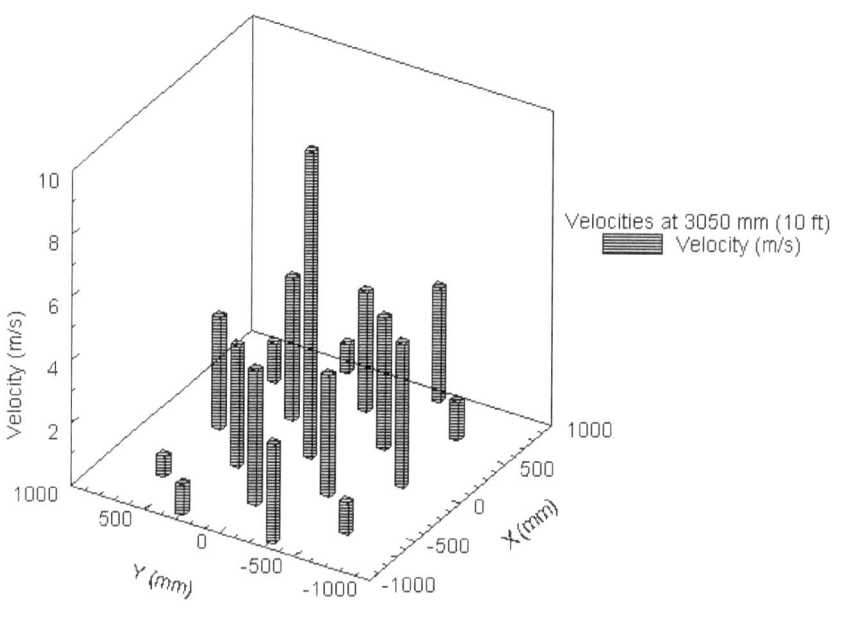

Figure 2-21. FDS Velocities 3.1 m (10 ft) From the Fan

Chapter 3: Simple Room Experiment

3.1 Experimental Description

A series of experiments were conducted to determine the impact that basic room geometries have on the flow of PPV fans. The experimental results were compared to FDS simulations to see if FDS predicted the airflow accurately.

3.1.1 Experimental Facility

These experiments were also conducted at NIST's Building and Fire Research Laboratory Large Fire Facility. The experiments were located in an area within the facility so that the airflow created by the PPV fan was not affected by external factors to the experiments. The facility had the following interior dimensions: 36.6 m (120 ft) long, 18.3 m (60 ft) wide and 7.6 m (25 ft) high.

3.1.2 Experiment Components

Fan and Anemometer
The same fan and anemometer used in the chapter 2 experiments were used in this series of experiments. A complete description of the fan and the anemometer can be found in section 2.1.2 of this report.

Room
The floor plan for the room is shown in Figure 3-1. The room was on a 0.2 m (8 in) high base with plywood decking and had a ceiling that was 2.6 m (8 ft - 8 in) high, measured from the top of the base. The window on the left hand side of the room was located 0.45 m (18 in) off the floor and was 1.4 m (54 in) tall. The door centered in the front wall was 2.0 m (80 in) tall. All of the walls were finished with gypsum board. There was a 1.8 m (6 ft) overhang that extended over the front of the room (Figures 3-2, 3-3 and 3-4).

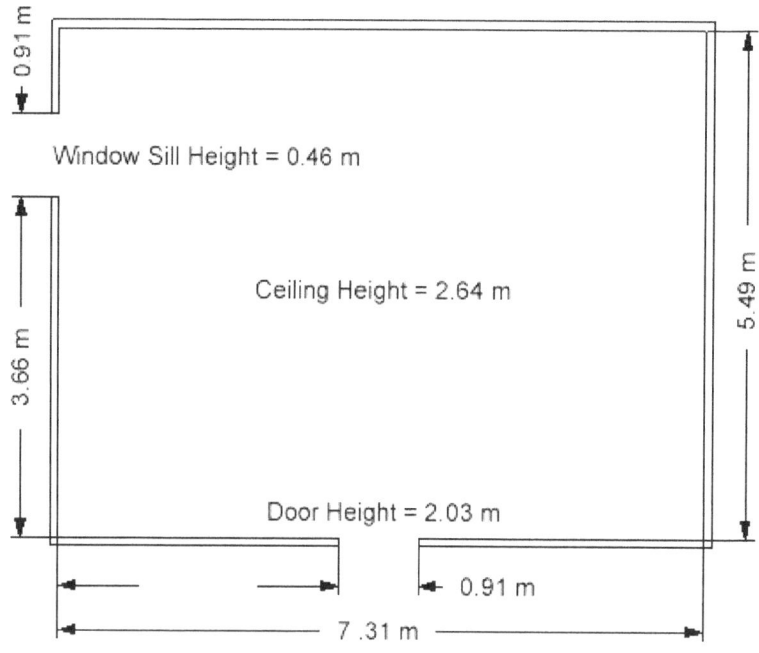

Figure 3-1. Floor Plan of Simple Room

3.1.3 Experiment Layout

The fan was positioned in the center of the doorway and allowed to run at the maximum speed, 2200 rpm. The cone of air coming from the fan covered the doorway when the fan was 3.05 m (10 ft) from the doorway so that was the location selected for the tests. The doorway and window were marked with anemometer measurement locations. These locations are shown in Figures 3-5 and 3-6.

Figure 3-2. Simple Room Experimental Layout

Figure 3-3. Experimental Layout Looking at Room Inlet and Outlet

Figure 3-4. Inlet and Outlet From a Different Perspective

3.1.4 Experiment Procedure

The fan was turned on and allowed to run for two minutes at the maximum speed setting, 2200 rpm. While the fan was running, the door was rechecked to make sure there was a cone of airflow around the door, and the window was checked to make sure there was a constant flow prior to measurement. Once the fan had been running for two minutes the data recording was started. Four minutes of readings were taken using sixteen second averaging with no rotation of the anemometer. Readings were recorded every two seconds and the output was an average of the previous eight readings. Air velocity readings were taken at specific points at the door and window as seen in Figures 3-5 and 3-6.

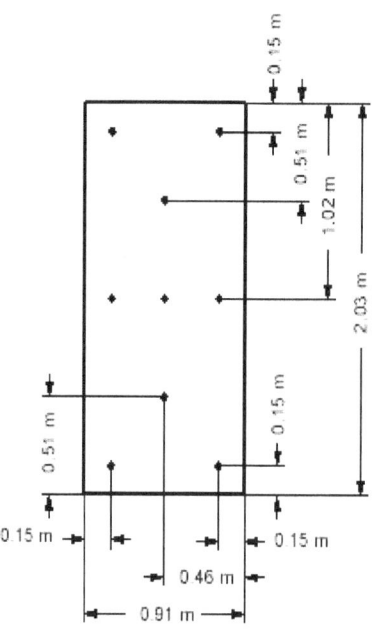

Figure 3-5. Doorway (Inlet) Measurements Points

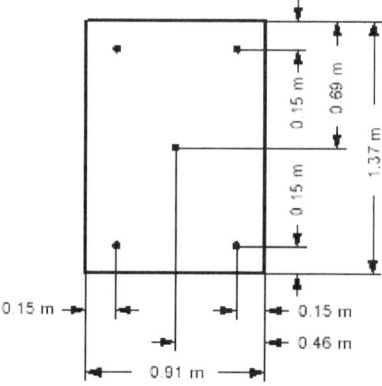

Figure 3-6. Window (Outlet) Measurement Points

3.1.5 Flow Visualization Experiment

In order to visualize and qualify the flow through the room a supplemental experiment was performed. A smoke generator was placed in the room and turned on to produce enough artificial smoke to fill the room. Cameras were set up and the fan was turned on to the maximum speed of 2200 rpm. Pictures were taken of the artificial smoke flowing from the room (figures 3-7 and 3-8).

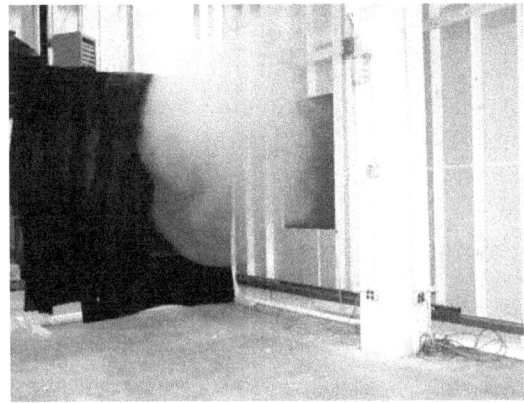 

Figure 3-7. Visualization of Supplemental Experiment as Soon as Room is Pressurized

Figure 3-8. Visualization of Supplemental Experiment Once Constant Flow is Achieved

3.2 Computer Simulation

The inputs required by FDS in order to model the PPV fan for this experiment were the same as those in chapter 2. The difference was the addition of the room. The room required additional inputs such as the properties of the ceiling, walls and openings.

3.2.1 Domain

The main issue that arises with the use of both the fan and the room was computational cell size. As stated in chapter 2 the simulated fan required a computation cell size of 16.4 cm^3 (1 in^3). If the domain containing the fan and room were completely 16.4 cm^3 (1 in^3) cells then there would be in excess of 15 million cells. Past experience with FDS suggests that if the number of cells exceeds one million, the computer computational time becomes excessive. In the case of 15 million cells one simulation would require over a month to complete on a Linux workstation.

In order to reduce the number of cells and still maintain sufficient accuracy of the calculation, FDS utilizes a multi-blocking feature. Multi-blocking allowed computational time to be saved by applying relatively fine grids in areas of interest and coarse grids elsewhere [6]. In this case, a fine grid was used for the domain surrounding the fan and a coarser grid was used for the room. There were many considerations that needed to be taken into account when using multi-blocking, especially for these cases with a large quantity of air movement. First, the finer grid containing the fan had to be specified first in the input file. This allowed FDS to give precedence to this domain and decreased calculation times. Second, the grids needed to overlap by at least 1.0 m. This helps accelerate calculations because information was transferred from grid to grid via external boundaries. When the grids overlapped they shared the same information, which also helped to speed up calculations. Third, neither of the domain boundaries could be within one meter of the face of the fan. This allowed for airflow in and out of the fan. Placing the end of a domain too close to the fan face caused inaccurate simulation of the air entrainment. Next, the fan grid could not be completely imbedded within the room grid. When this was done information is not transferred between the grids and there would be no results for the fan. Finally, the cell size for the room domain could not be larger than 0.1 m (4 in) on a side. This allowed for a small step between the grids sizes and sufficient amounts of data to be shared.

Another issue that needs to be addressed was the positioning of the PPV fan. From the Smokeview visualization, it appeared that the fan was floating in mid air. The reason for this was the inability of FDS to portray angled geometry. Typically the fire service places the fan away from the door and adjusts the angle of the fan to achieve a cone surrounding the opening [9]. In order to reduce the complexity of the simulation the fan was modeled at the mid-height of the doorway. The fluid dynamics of the two situations are assumed to be similar.

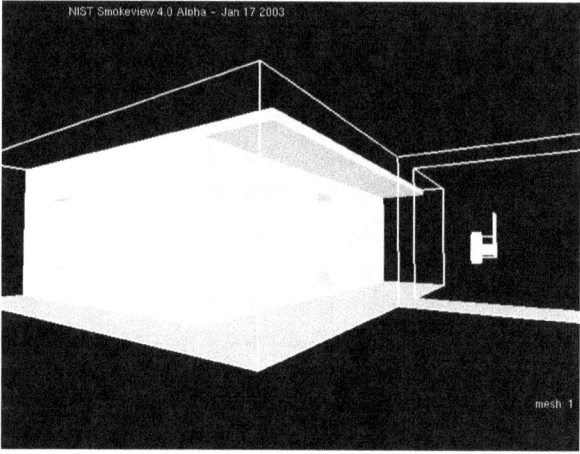

Figure 3-9. FDS Layout for Simple Room Experiment

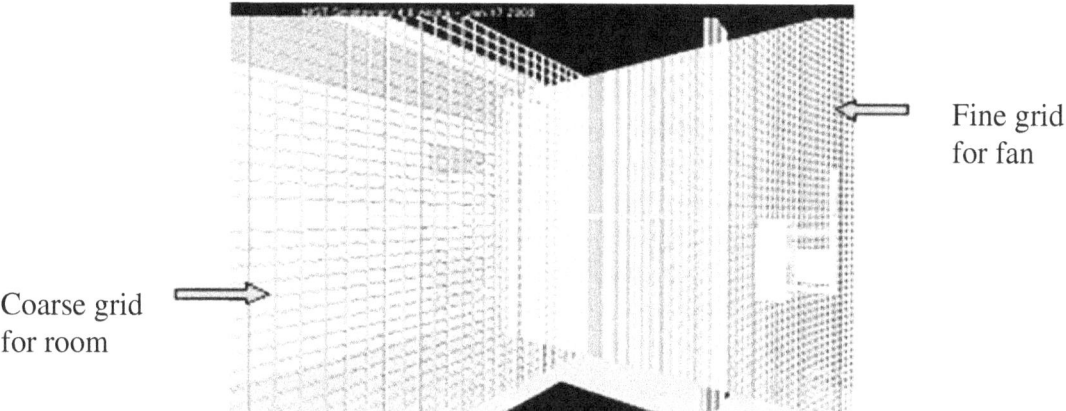

Figure 3-10. Grid Cell Visualization, Multi-blocking

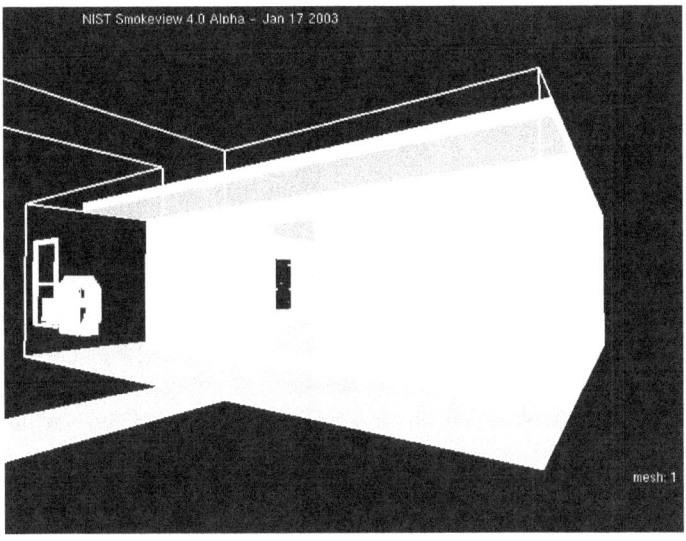

Figure 3-11. FDS Layout Looking at Room Inlet and Outlet

3.2.2 Geometry and Vents

The PPV fan used for these runs was the same fan created in Chapter 2. See section 2.2.3 for details on fan geometry and vent considerations. Chapter 3 involved the addition of a room with dimensions of that in Figures 3-1 through 3-5. This room was prescribed in FDS as a series of blocks with characteristics of gypsum board. The dimensions and openings were the same as those in the experimental setup. All of the room boundaries, including the walls and ceiling were contained within the domain.

3.2.3 Output Files

Vertical velocity slices were positioned through the center of the doorway and window. Velocity measurement points were also prescribed at the same locations as in the experimental layout. These output files were compared with the experimental measurements (Section 3.3).

3.3 Results

Velocity measurements were taken at the locations in the door shown in Figure 3-5, and in the window in Figure 3-6. These measurements were recorded with the fan 3.0 m (10 ft) from the front door of the room, just as in the chapter 2 experiments. Having the fan 3.0 m (10 ft) from the door provides a cone of air around the front door per International Fire Service Training Association recommendations [9]. The magnitudes of the velocities into the door are shown in Figure 3-13. The magnitudes of the outlet velocities at the window are in Figure 3-15. The average inlet velocity at the door was 2.2 m/s (7.2 ft/s). The average outlet velocity at the window was

2.6 m/s (8.5 ft/s). The model was used to examine the use of the PPV fan from chapter 2 in the same configuration as in the experiment. FDS yielded an inlet velocity averaged over the door area of 4.1 m/s (13.5 ft/s) and an outlet velocity of 3.0 m/s (9.8 ft/s) averaged over the window area (Figures 3-14 and 3-16). These comparisons provided a difference of 84 % at the door and a difference of 16.5 % at the window (Table 2). The air movement at the window was the result of the pressure change in the room, interaction of the geometry of the room and the airflow. The door point predictions were dependent upon the local turbulence that FDS creates from the fan simulation shown in Figure 3-12. Experimentally this turbulence was very complex and difficult to measure. Qualitatively, the flow was compared in Figures 3-19 to 3-22.

Table 2. Comparison of Experimental and FDS Simple Room Results

	Average Velocities, m/s (ft/s)	
Door		Window
Experimental	2.2 (7.2)	2.6 (8.5)
Fire dynamics simulator	4.1 (13.5)	3.0 (9.8)
Relative Difference (%) (Exp-FDS)/Exp x 100	84.0	16.5

Note: Experimental velocity total expanded uncertainty is ± 14%, See Table 10.

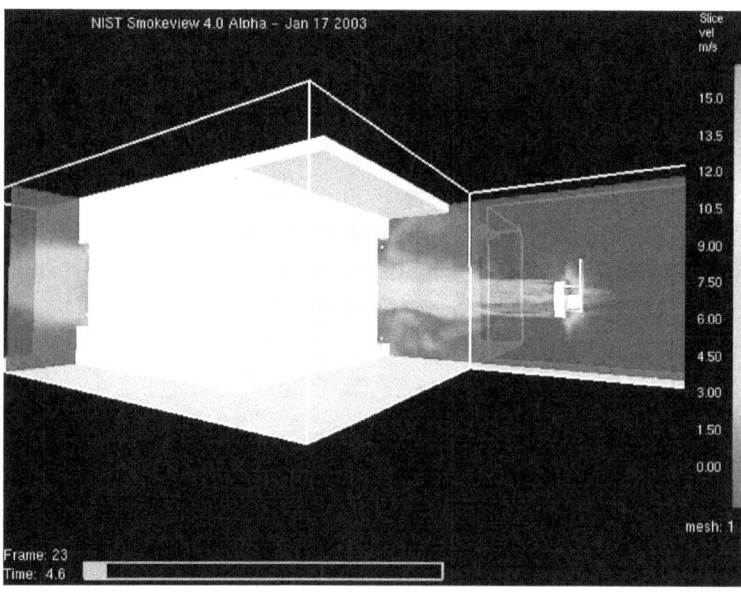

Figure 3-12. FDS Layout with Fan Operating

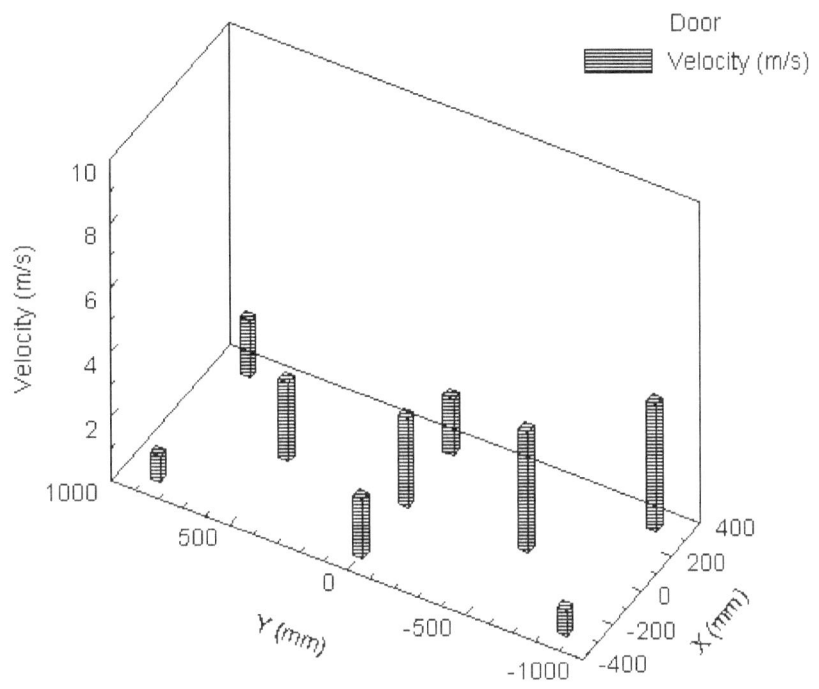

Figure 3-13. Experimental Velocity Measurements in the Doorway (Inlet)

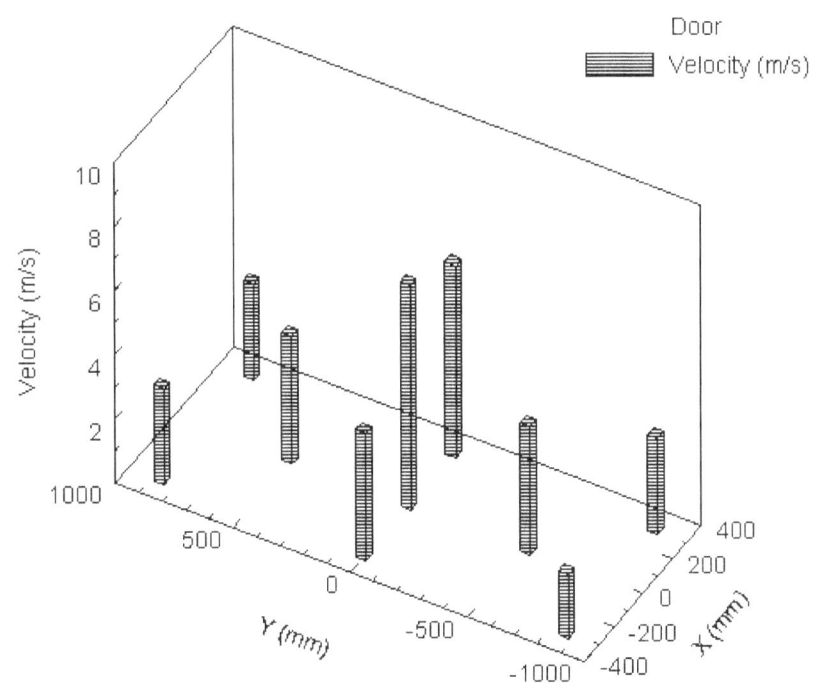

Figure 3-14. FDS Velocities in the Doorway (Inlet)

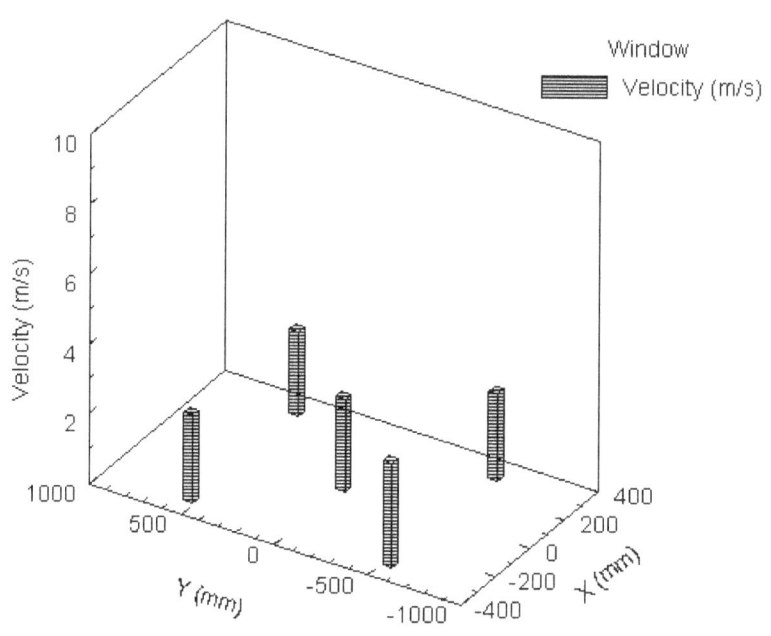

Figure 3-15. Experimental Measurements in the Window (Outlet)

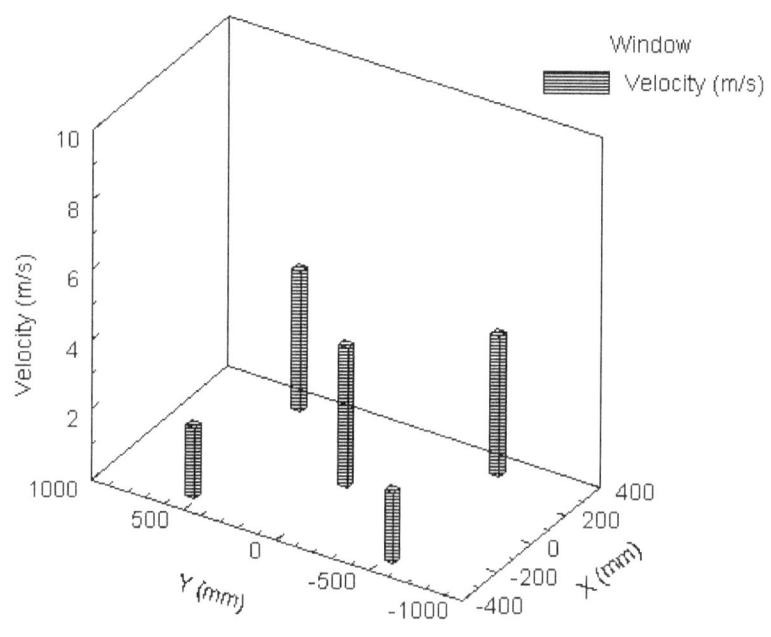

Figure 3-16. FDS Velocities in the Window (Outlet)

3.4 Mapping and Simple Room Summary

A comparison of the computational fluid dynamic model Fire dynamics simulator (FDS) was made with data from two different sets of data collected from full-scale experiments. The full-scale experiments characterized a Positive Pressure Ventilation (PPV) fan in an open atmosphere, and with a simple room geometry. Both experimental data sets provided insight into the gas velocities, as well as providing the opportunity to validate the predictions of the Fire dynamics simulator.

The measurements for the fan in an open atmosphere compared favorably with the FDS predictions. With the correct geometry, vent placement and boundary location FDS predicted velocities that were within 10 % of the experimental results (Table 3). FDS's visualization of the flow pattern also correlated well with the experimental visualization.

The measurements for the fan and a single room also compared favorably with the FDS predictions for the flow out the window. The flow that was created out of the window in FDS was within 20 % of that measured experimentally. FDS's visualization of the flow out of the window using Multi-blocking also correlated well with that captured experimentally (Table 4).

The results from these two experiments indicated that FDS was able to portray the general room flow created by PPV fans. Future experiments need to be conducted to examine the flow in multi-floor structures and structures with a more complex geometry. Further, the impact of the fan on other factors in the fire environment such as temperature, burning rate, and gas concentrations has not been examined here.

Table 3. Summary of FDS Characteristics Needed to Model the Fan Flow

1	Computational cell size of less than or equal to 1 cubic inch
2	At least 1 meter of domain to the rear of the fan
3	Shroud must be cylindrical
4	Vents must be located on the front face of the shroud
5	A velocity must be prescribed for the vents (17.88 m/s for this fan)
6	A block must be added to the center of the shroud to simulate the blade center (effects air movement)
7	A motor must be added to the rear of the fan (effects air movement)

Table 4. Summary of Multi-block Characteristics Needed to Model Simple Room Flow

1	Specify the fan grid (finer) first in the data file
2	Overlap the grids by at least 1 meter
3	Neither of the domain boundaries should be within 1 meter of the fan
4	Do not embed one grid completely within the other
5	Computational cell size for the non-fan grid should not be greater than 102 mm (4 in) on a side.

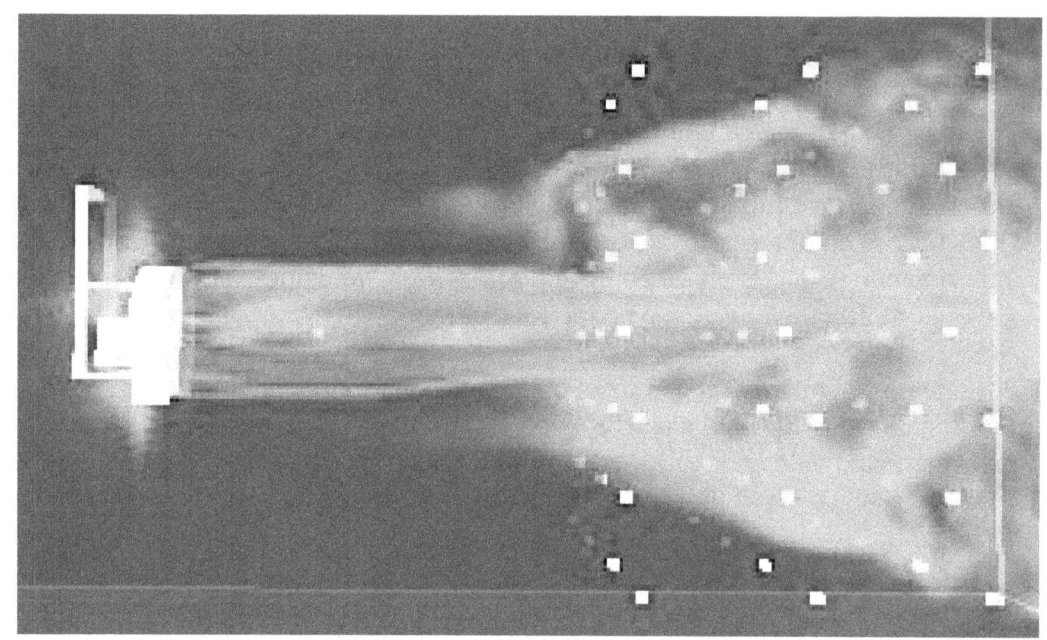

Figure 3-17. Smokeview Visualization of FDS PPV Flow Pattern, see Figure 2-15 for velocities

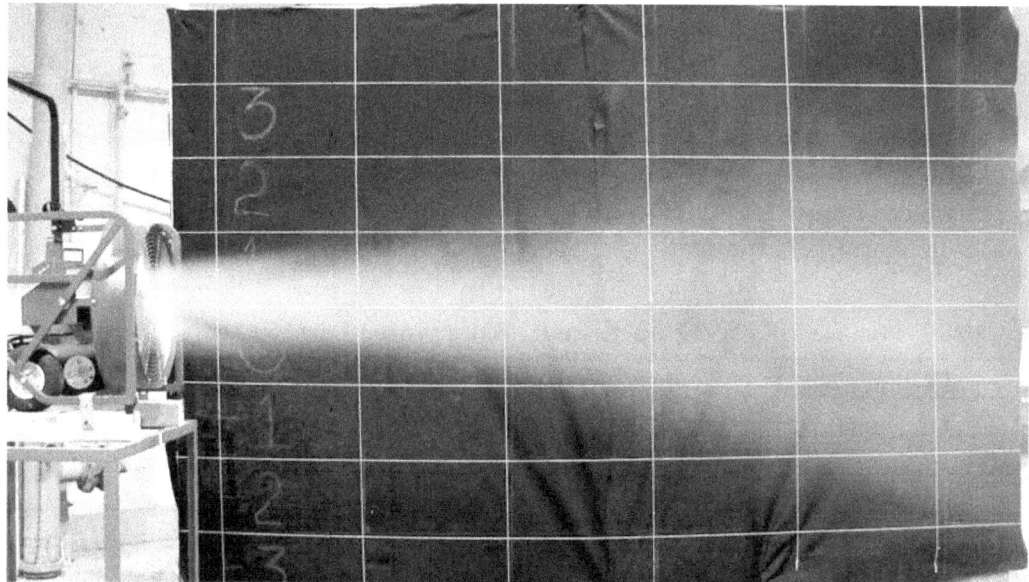

Figure 3-18. Experimental Visualization of the PPV Flow Pattern

Figure 3-19. FDS Visualization of Supplemental Experiment as Soon as Room is Pressurized, see Figure 3-12 for velocity magnitudes

Figure 3-20. Visualization of Supplemental Experiment as Soon as Room is Pressurized

Figure 3-21. FDS Visualization of Supplemental Experiment Once Constant Flow is Achieved, see Figure 3-12 for velocity magnitudes

Figure 3-22. Visualization of Supplemental Experiment Once Constant Flow is Achieved

Chapter 4: Room Fire Experiments

4.1 Experimental Description

A pair of full-scale experiments was performed at NIST's Building and Fire Research Laboratory Large Fire Facility. The facility has interior dimensions, 36.6 m (120.0 ft) long, 18.3 m (60.0 ft) wide and 7.6 m (25.0 ft) high. A room was constructed within the facility under a 6 m x 6 m (20 ft x 20 ft) hood which was instrumented to allow for oxygen depletion calorimetry to measure the heat release rate produced by the fire in the room. The room had interior dimensions of 3.66 m x 4.27 m (12 ft x 14 ft) and a ceiling height of 2.44 m (8 ft) (Figure 4-1). A window was located in the center of one of the 4.27 m walls (Figure 4-2) and a doorway was located on the center of one of the 3.66 m walls. The doorway opened to a 1.22 m wide x 2.29 m long (4.0 ft x 7.5 ft) corridor (Figure 4-3). Another doorway was located at the end of the corridor that was the same size as the doorway to the room. Both doorways were 2.06 m (6.75 ft) tall and 0.91 m (3 ft) wide. The window was 1.2 m (3.9 ft) tall and 0.89 m (2.9 ft) wide and 0.81 m (2.7 ft) from the floor to the sill (Figure 4-1). A 6.1 m (20 ft) wall was constructed between the corridor doorway and room window to the exterior of the room in order to isolate the effects of the fan and not allow recirculation of smoke from the window back through the doorway. All of the walls and the ceiling were framed with 0.050 m x 0.100 m (2.0 in x 4.0 in) pine studs and sheathed with two layers of 0.0095 m (0.38 in) gypsum board. The room was furnished with a bunk bed, stuffed chair, book case and desk (Figure 4-4). The floor was covered with carpet and a computer monitor was placed on the desk.

Two experiments were conducted with nearly identical fuel loads, examining the effect of the PPV fan on the room fire. The first experiment utilized the same fan used in previous experiments to forcibly ventilate the room just after the window was opened. The second experiment was similar to the first experiment except that it was naturally ventilated. The fan used was identical to that used in the previously described experiments. It was a 0.75 m (18 in) in diameter and had a depth of 0.48 m, (18.75 in), width of 0.62 m (24.5 in) and height of 0.62 m (24.5 in). It had a maximum speed of 2200 rpm, a power rating of 746 W (1 hp) and a volumetric flow rating of 6.64 m^3/s (14,060 ft^3/min) [10]. The fan was positioned 2.44 m (8 ft) from the open doorway to the corridor at an angle of approximately 15 degrees from horizontal to create the cone of air around the doorway.

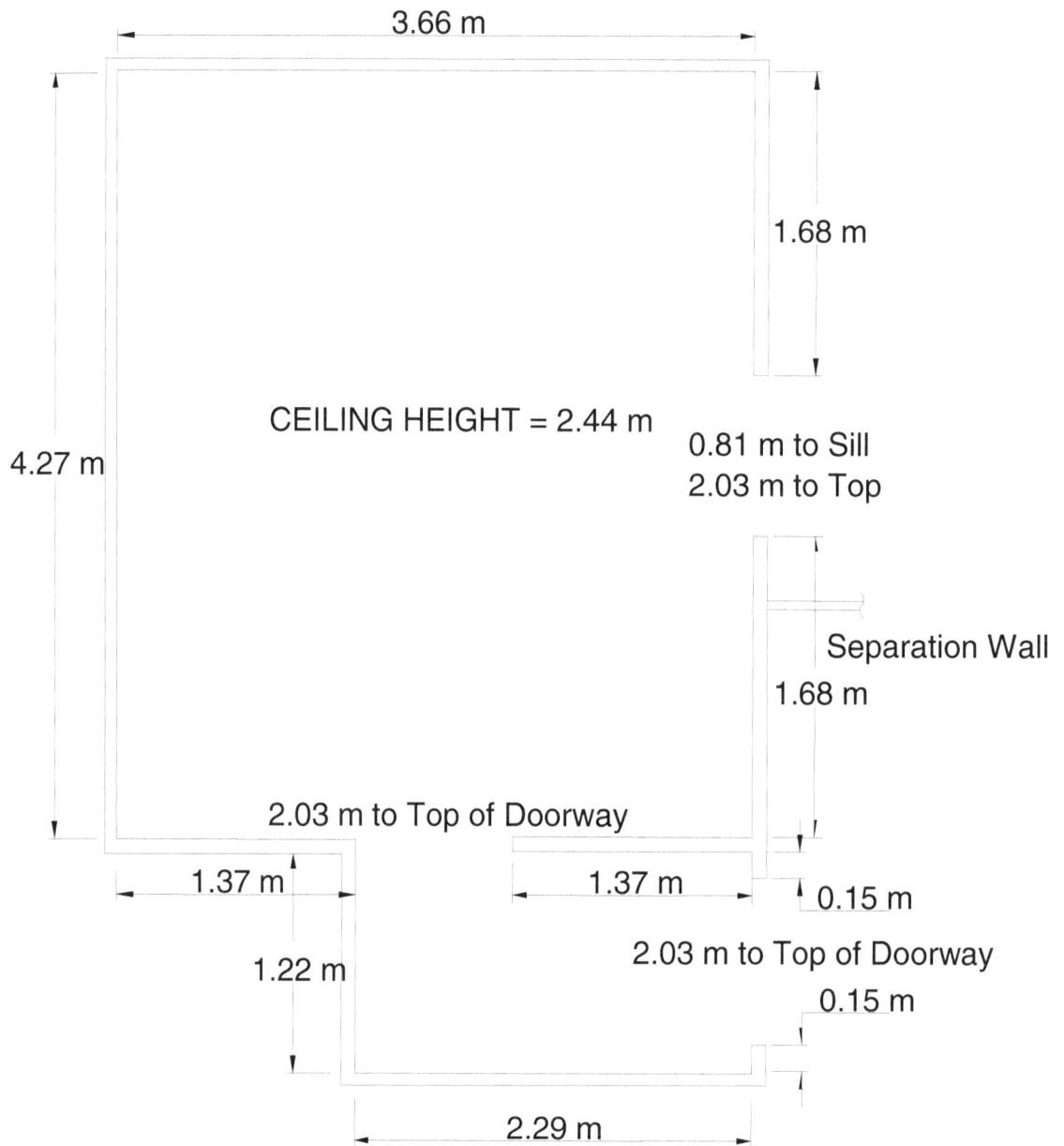

Figure 4-1. Experimental Floor Plan

Figure 4-2. External View of Window in Closed Position

Figure 4-3. External View of Open Corridor Doorway

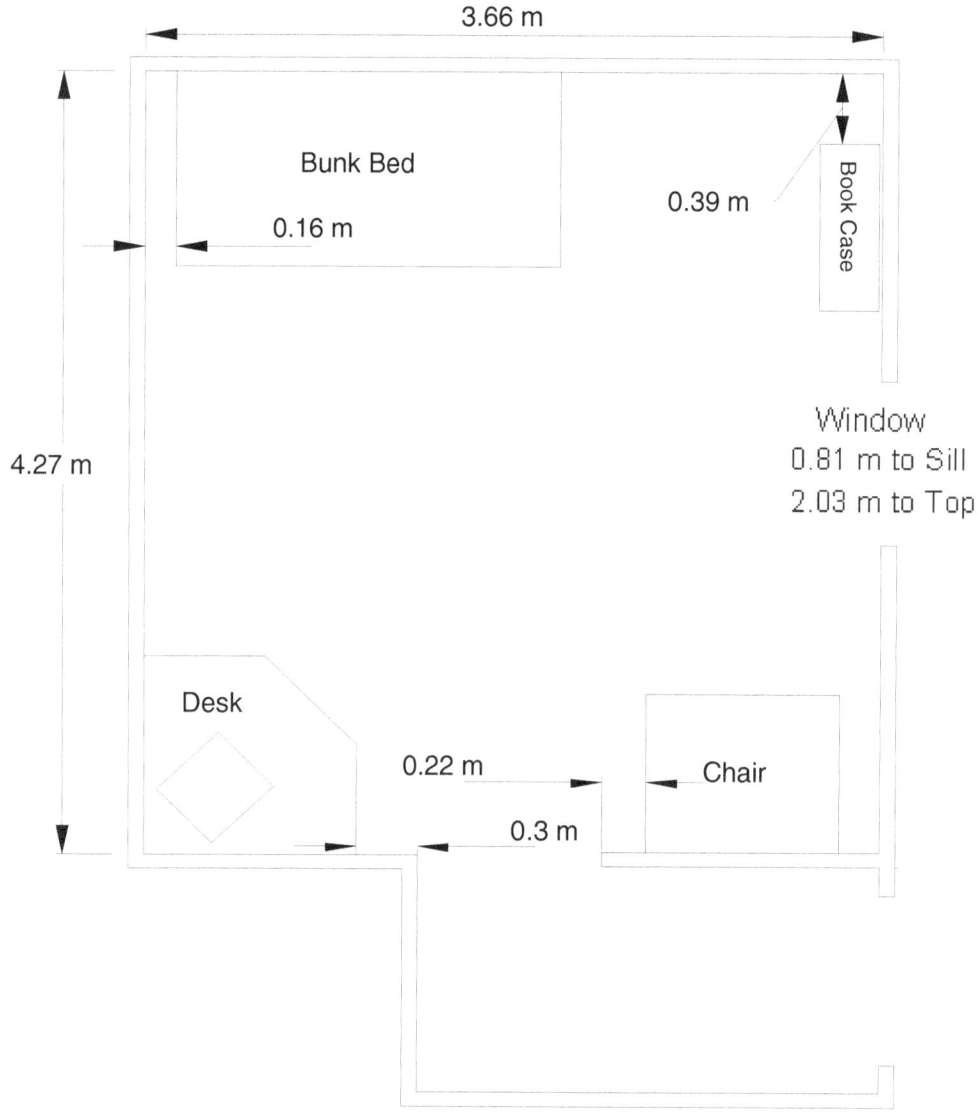

Figure 4-4. Furniture Floor Plan

4.1.1 Instrumentation

Temperature measurements were made with 0.5 mm (0.02 in) nominal diameter type K bare bead thermocouples. A vertical thermocouple array was located in the center of the fire room with measurement locations of 0.025 m, 0.30 m, 0.61 m, 0.91 m, 1.22 m, 1.52 m, 1.83 m and 2.13 m (1 in, 1 ft, 2 ft, 3 ft, 4 ft, 5 ft, 6 ft and 7 ft) below the ceiling. The corridor doorway also had a vertical array with measurement locations of 0.025 m, 0.30 m, 0.61 m, 0.91 m, 1.22 m, 1.52 m, 1.83 m (1 in, 1 ft, 2 ft, 3 ft, 4 ft, 5 ft and 6 ft) below the top of the doorway opening. Three thermocouples were located in the room doorway, 0.30 m (1 ft) from the top of the doorway, 0.30 m (1 ft) from the bottom of the doorway and the midpoint of the doorway at 1.02 m (Figure 4-5). Six additional thermocouples were placed in the ventilation window.

Three were located 0.30 m (1 ft) in from each side of the window at heights of 0.15 m, 0.61 m and 1.07 m (0.5 ft, 2 ft and 3.5 ft) from the bottom of the window (Figures 4-6 and 4-7).

Gas velocity measurements were recorded in the window and in the fire room doorway using bi-directional probes (Figure 4-8). The doorway had measurement locations of 0.30 m (1 ft) from the top of the doorway, 0.30 m (1 ft) from the bottom of the doorway and the midpoint of the doorway. The ventilation window had six bi-directional probes in the same locations as the thermocouple locations shown in Figure 4-6. Probes were connected to Setra Systems Model 264 differential pressure transducers. Since both in and outflow was expected, each transducer had the capacity to monitor positive and negative pressure differentials. Bi-directional high pressure ranges up to 62 Pa (0.01 psi) were utilized because of the flow that was expected in the two openings. A set of pressure transducers was also positioned in the rear corner of the room adjacent to the bookcase to examine the differential pressure created at 0.30 m, 1.22 m, 2.13 m (1 ft, 4 ft and 7 ft) from the floor [11].

Video recordings were made of each experiment. Two cameras were positioned on the exterior of the room. One had the view of the ventilation window and the other the open doorway to the corridor. Two water cooled cameras were positioned inside the room (Figure 4-9). One camera was placed just above the floor near the bunk beds viewing the gas flow between the window and doorway. The second interior camera was positioned near the floor viewing the ignition location and flame spread across the bunk beds.

The combustion products were captured by a 6 m x 6 m (20 ft x 20 ft) hood which was instrumented for oxygen consumption calorimetry. Oxygen depletion, temperature and pressure readings were continuously monitored in order to calculate the total heat release rate of the fire room [12].

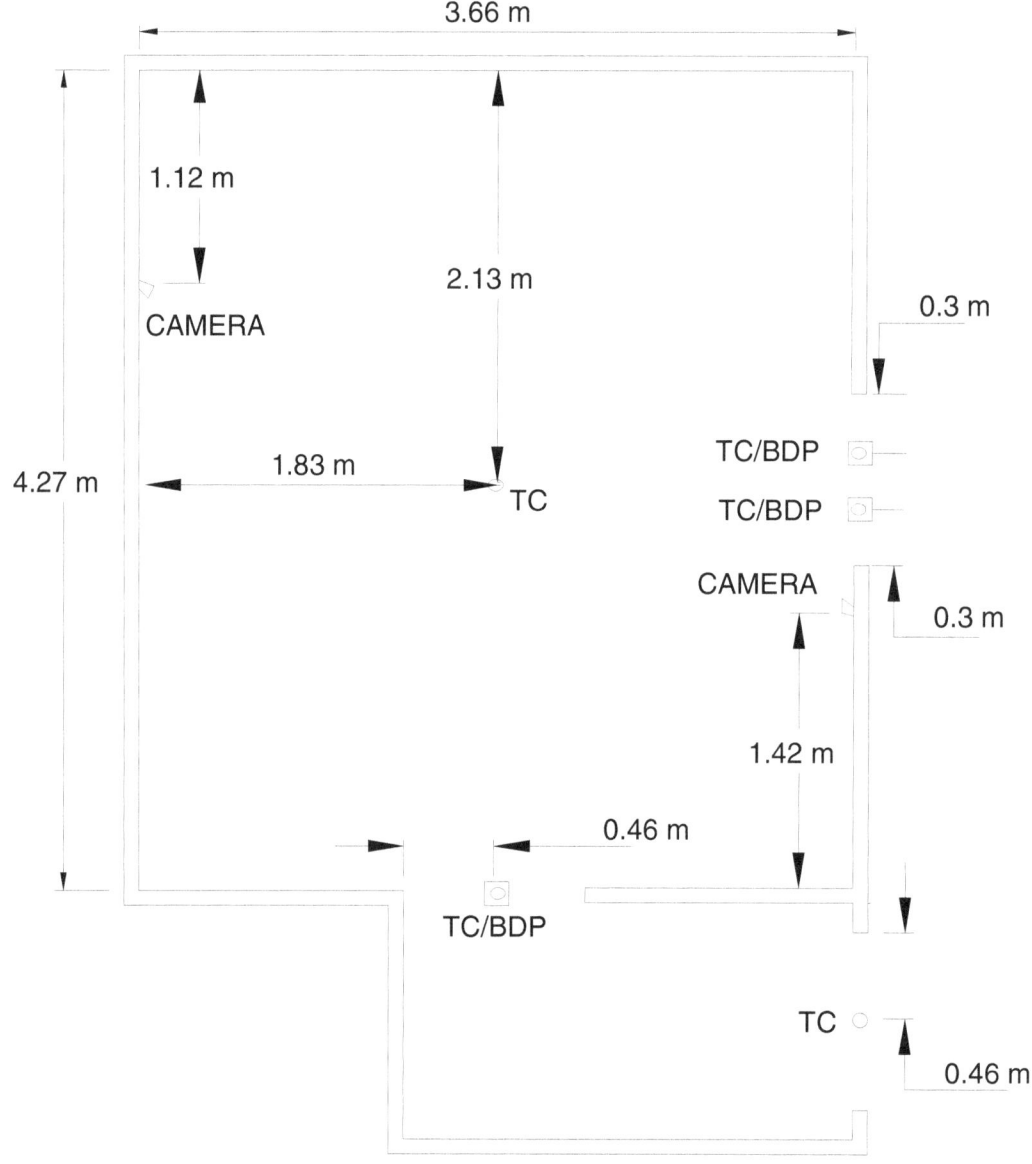

Figure 4-5. Instrumentation Floor Plan (TC: thermocouples, BDP: bidirectional probes)

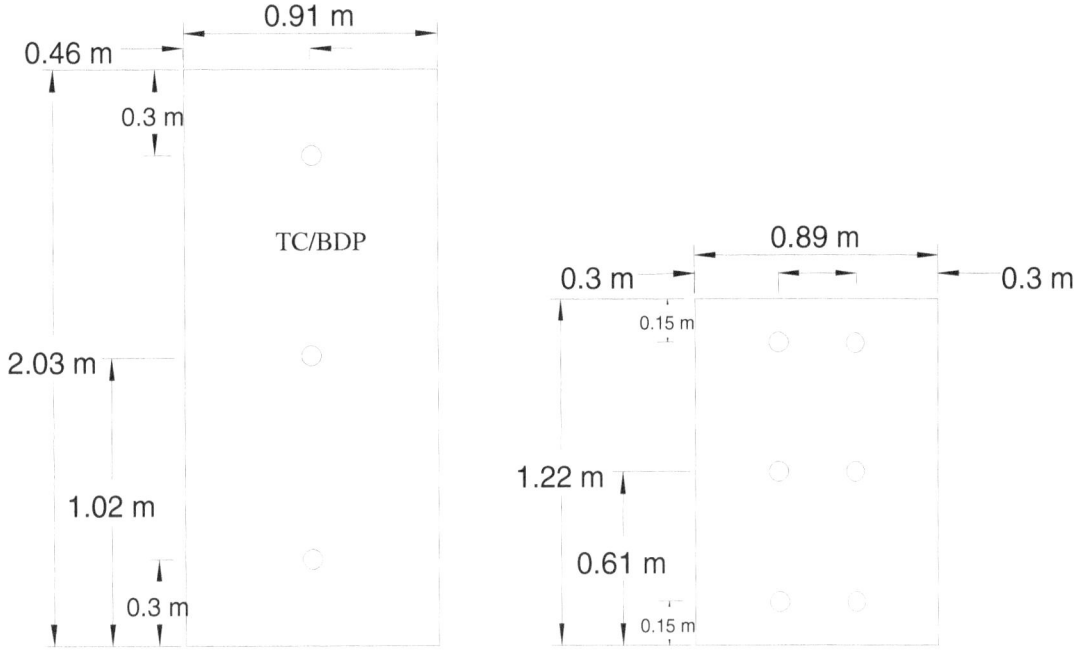

Figure 4-6. Doorway and Window Probe and Thermocouple Combination

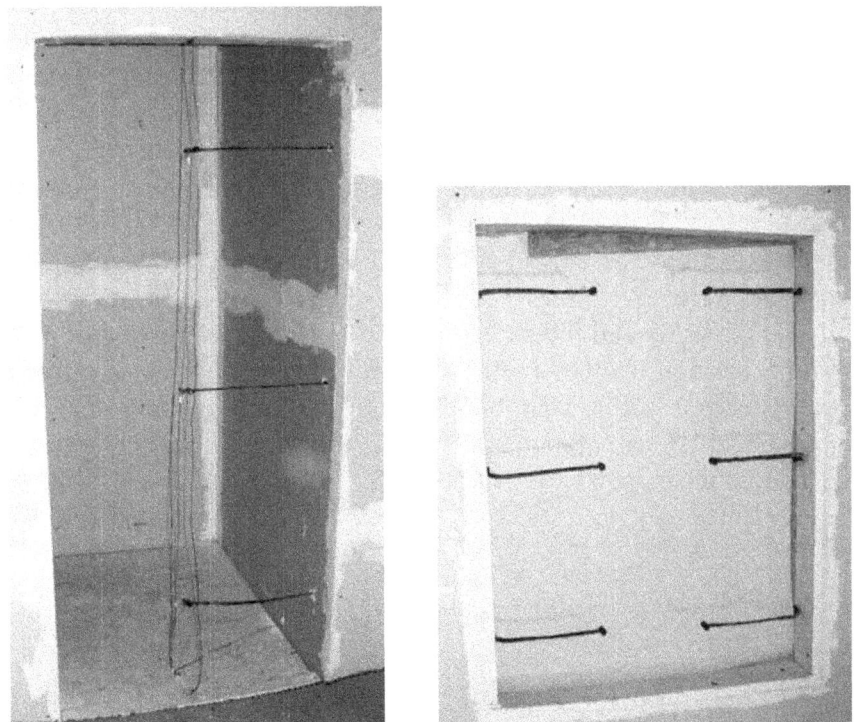

Figure 4-7. View of Doorway From Inside the Room into Corridor and Window with Instrumentation

Figure 4-8. Bidirectional Probe and Thermocouple Combination   **Figure 4-9. Water Cooled Camera**

4.1.2 Fuel Load

The fuel load was selected to represent a typical child's bedroom. It was also intended to create a fuel rich atmosphere to make burning dependent on the available oxygen. Both experiments had similar furniture and a total fuel mass that was 250.6 kg and 251.6 kg for experiment 1 and 2, respectively (Table 5). The book case was made of 0.013 m (0.5 in) thick compressed particle board covered with a plastic laminate top (Figures 4-10 and 4-11). The desk was also made of compressed particle board with a laminate top, but was 0.038 m (1.5 in) thick (Figures 4-12 and 4-13). The 0.38 m (15 in) nominal size computer monitor on the desk had a plastic shell and a glass face (Figures 4-14 and 4-15). The chair located in the corner had a wood frame, polyurethane cushions and cotton cover (Figures 4-16 and 4-17). Finally, the bunk beds were framed out with 0.050 m x 0.100 m (2 in x 4 in) nominal pine lumber (Figures 4-18 and 4-19). All of the mattresses were placed on box springs and were covered with cotton bedding consisting of fitted sheets, blankets, comforters, pillows and pillow cases. The mattresses consisted of a polyester cover, twin inner springs, and polyurethane foam. The carpet was polypropylene based and covered the floor wall to wall, but did not extend into the corridor. There was no padding under the carpet, but a 0.01 m (0.375 in) thick sheet of gypsum board was placed under the carpet.

Table 5. Fire Load Weights

Item	Mass (PPV Experiment) (kg)	Mass (Natural Experiment) (kg)
Bunk Bed Frame	19.15	20.13
Top Mattress	14.60	17.14
Top Box Spring	12.63	13.06
Bottom Mattress	14.98	14.44
Bottom Box Spring	14.21	12.94
Pillows	0.57 0.56 / 0.58 0.56	
Pillow Cases	0.11 0.10 / 0.11 0.10	
Fitted Sheets	0.40 0.34 / 0.39 0.37	
Blankets	1.15 1.19 / 1.18 1.22	
Comforters	1.54 1.55 / 1.59 1.56	
Chair	27.13	27.80
Book Case	27.08	26.17
Desk	55.48	55.48
Computer Monitor	17.09	17.90
Carpet	40.66	39.00
Total Weight	250.63	251.61

Note: Experimental mass measurement total expanded uncertainty is ± 12%, See Table 11.

Figure 4-10 - Book Case

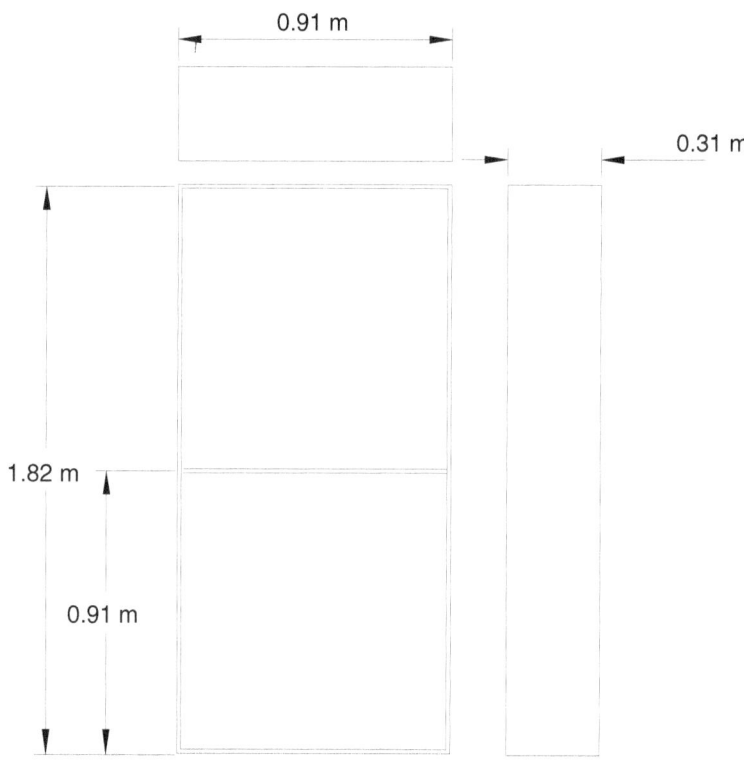

Figure 4-11 - Book Case Dimensions

Figure 4-12 - Desk and Monitor

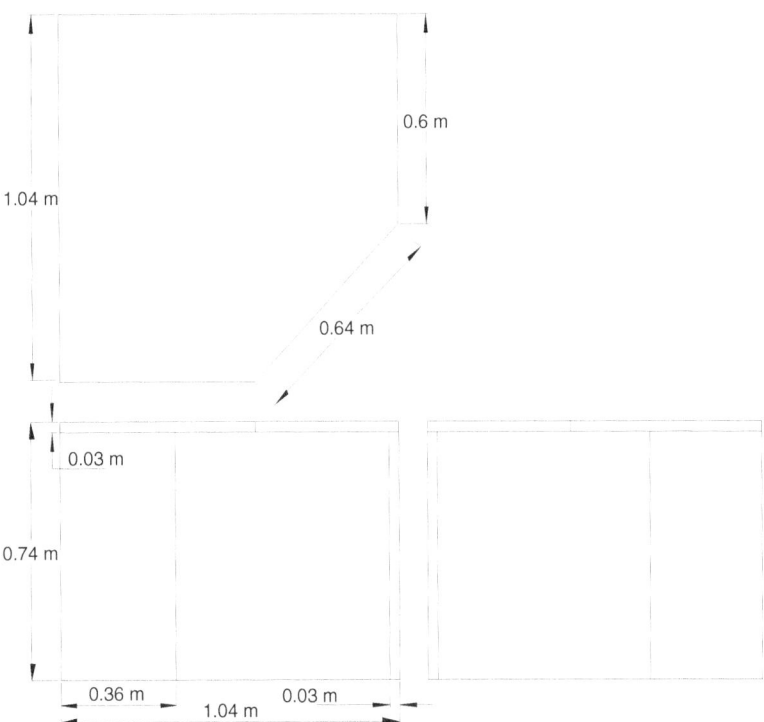

Figure 4-13 - Desk Dimensions

Figure 4-14 - Monitor

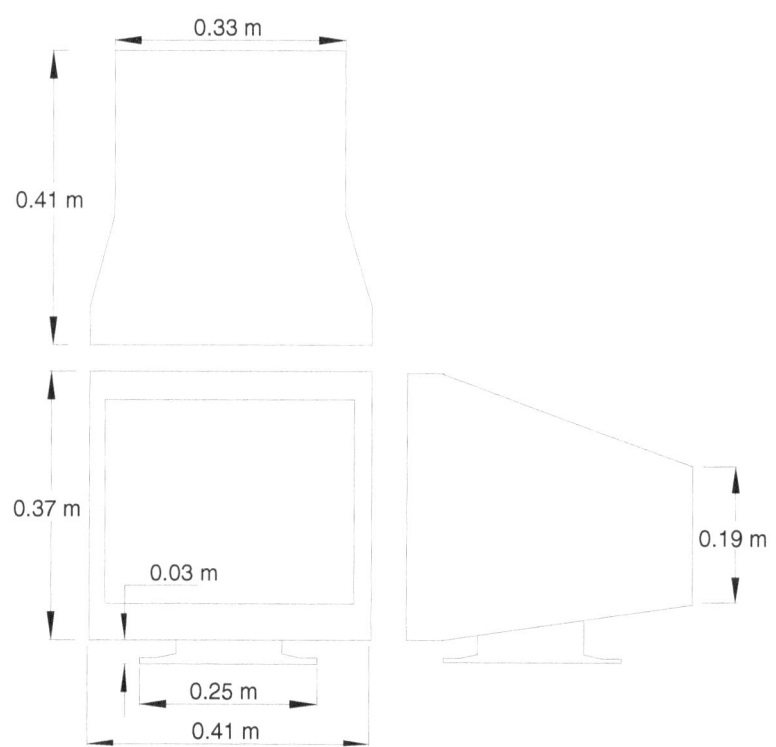

Figure 4-15 - Monitor Dimensions

Figure 4-16 - Chair

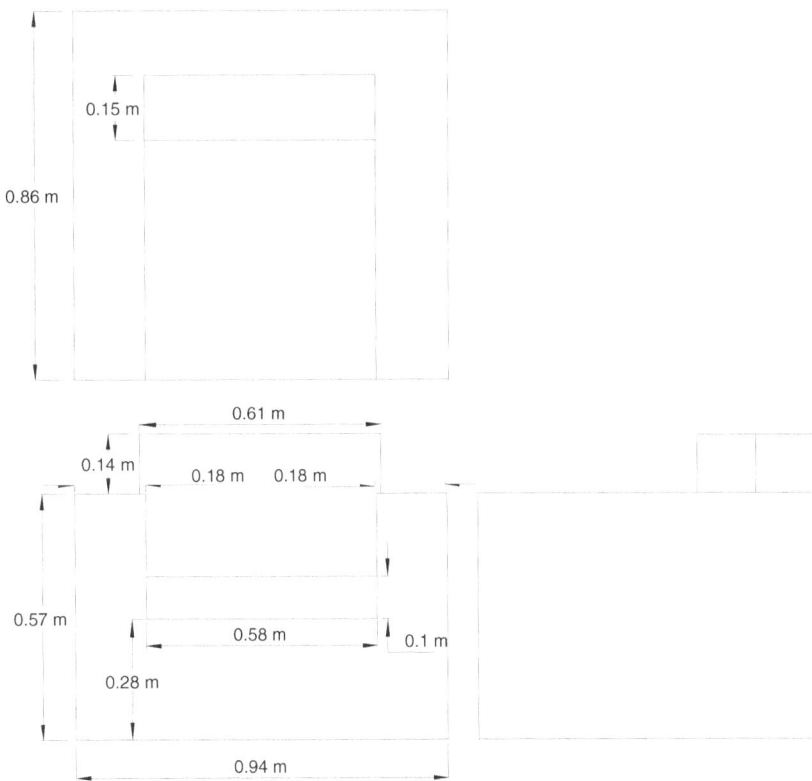

Figure 4-17 - Chair Dimensions

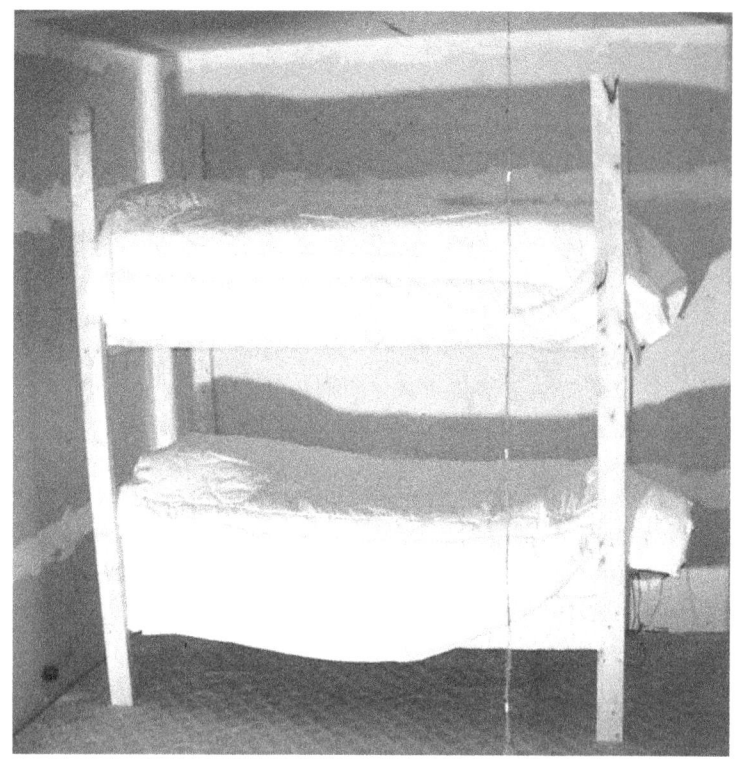

Figure 4-18 – Bunk Bed

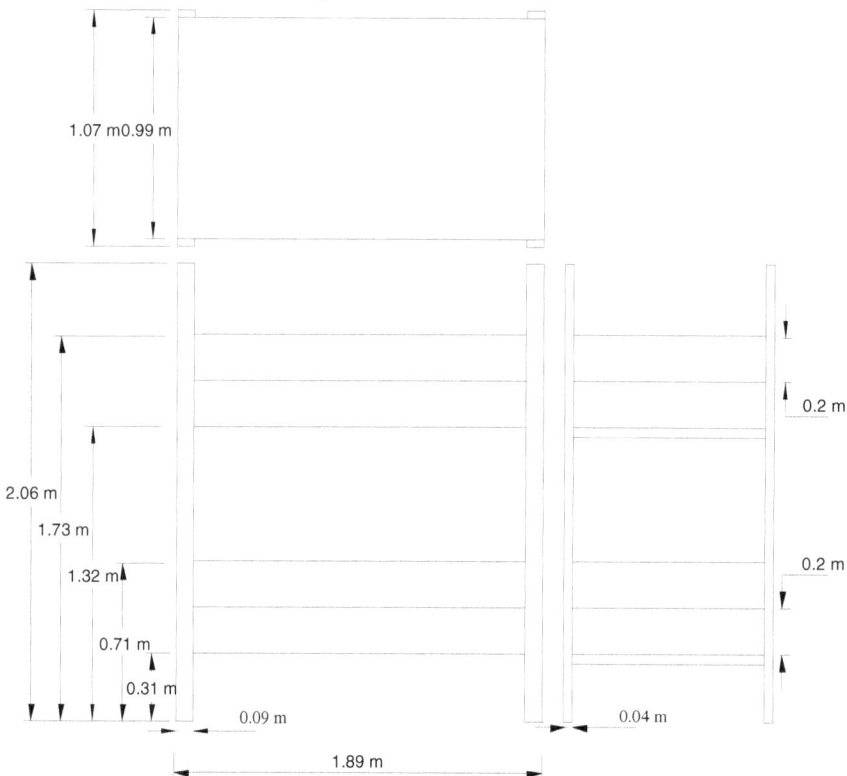

Figure 4-19 – Bunk Bed Dimensions

4.1.3 Procedure

The fire was ignited using two electrically activated matches located in the mattress of the lower bunk, in the corner nearest the center of the room (Figure 4-20). The electric matches consisted of a match book with the cover placed behind the matches, exposing the match heads. Nickel-Chromium wire was spiraled through the match heads and taped to the bottom corners of the match book. A copper wire was connected to each end of the nickel-chromium wire with alligator clips and run to the exterior of the room where they were connected to an igniter box. The igniter box sent current through the wire, heating the wire and, in turn, igniting the matches. One of the match books was cut into the mattress and the second was placed under the bedding, directly on top of the mattress to ensure a strong ignition (Figure 4-21).

At the time of ignition, the window was closed and the fire's only source of oxygen beyond the room was through the open room doorway which connected to the corridor via a doorway. The fire was allowed to grow until flashover conditions were reached and the fire became oxygen limited. This was determined by the internal video. Once the fire was oxygen limited for a short period of time, the window was opened from the outside of the room to ventilate the fire. Both of the experiments were ventilated 345 s after ignition. In the PPV experiment, the window was opened and 5 s later the fan was turned on to full speed until 1380 s after ignition, when the fan was turned off to assess the structure and begin extinguishment. In the naturally ventilated experiment, the window and doorway provided ventilation until the fuel in the room burned to completion. Table 6 provides the timeline of events.

Table 6. Experimental Procedure

Time (s)	Natural Ventilation	Positive Pressure Ventilation
0	Ignition	Ignition
345	Window Open	Window Open
350	-	PPV Fan On
* Fire burned until all fuel was exhausted		

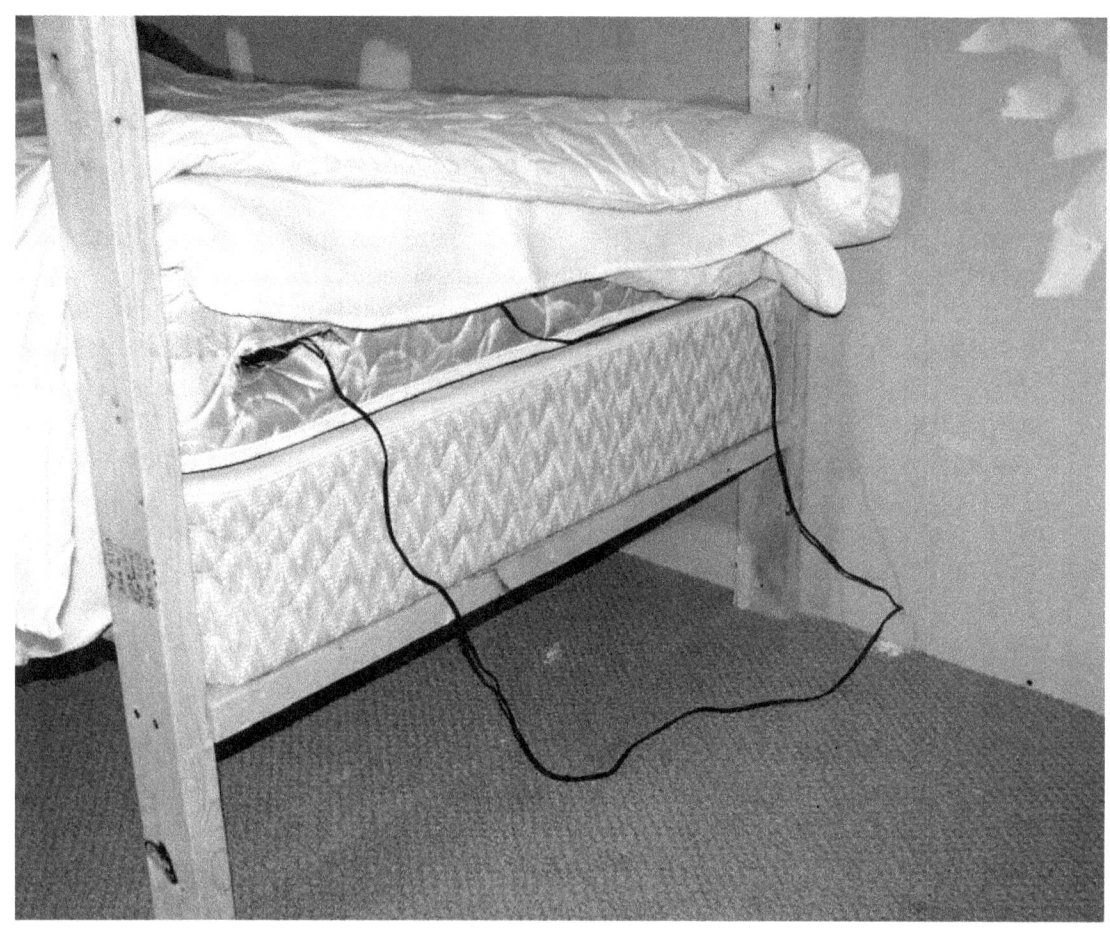

Figure 4-20. Ignition Setup

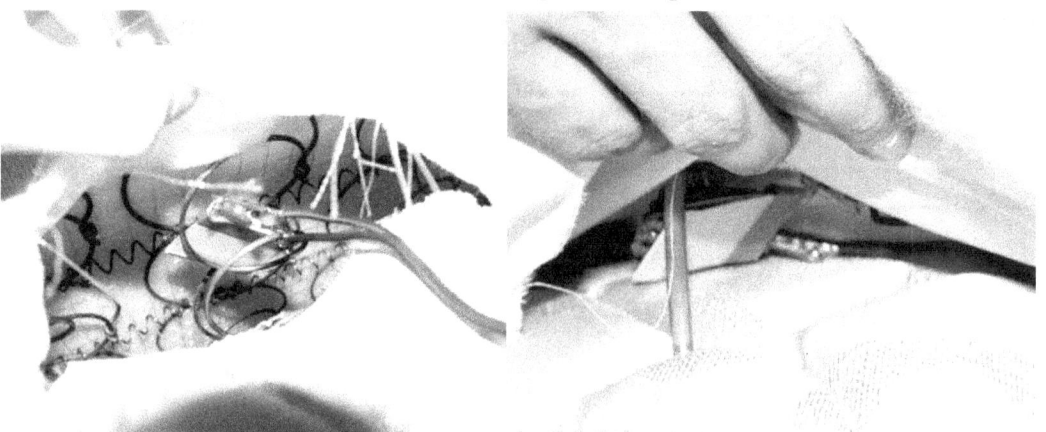

Figure 4-21. Electrically Activated Matchbook Locations

4.2 Experimental Results

Cameras at four locations allowed for the visualization of most of the fire growth and the combustion gas flow out of the furnished room. Both experiments experienced very similar fire growth up until the time that the window was opened. In the PPV ventilated experiment, flashover occurred at 275 s and near zero visibility at 278 s. In the natural ventilation experiment, flashover occurred at approximately 285 s and near zero visibility at 298 s. While the growth was very similar in both experiments, visibility returned more rapidly in the PPV experiment. The view from the water cooled camera focused on the corner of the room with the bunk beds. The camera showed visibility began to return 181 s after the ventilation in the PPV experiment and 395 s after the ventilation of the naturally ventilated case. Clear visibility inside the room returned to the PPV ventilated experiment 120 s prior to that of the naturally ventilated experiment.

In both experiments, black smoke was observed in the corridor prior to 300 s and flames were not observed in the corridor doorway until the window was ventilated. Within 10 s of opening the window, flames extended out of the corridor doorway. Once the fan was activated, it required 130 s to reverse completely the flow back into the room. The PPV fan forced all of the smoke and flames out of the corridor and back into the room by 516 s after ignition. At that point, little or no smoke was seen coming out of the room doorway. Flames were observed in the corridor of the naturally ventilated experiment until 1200 s (Figures 4-22 and 4-24).

The exterior view of the window showed that it took less than 5 s for flames to come out of the window after ventilation in both experiments. The flames in the PPV ventilated experiment extended approximately 1.83 m (6 ft) from the window. This length was referenced by the known width of the gypsum board sheets in the background. The flames from the naturally ventilated experiment extended approximately 0.91 m (3 ft) from the outside edge of the window (Figures 4-23 and 4-25). Detailed observations of both experiments are tabulated in Table 7.

Table 7. Observations

PPV Ventilated Experiment Time (s)	Natural Ventilation Experiment Time (s)	Observation
0 0		Ignition
86	120	Flames touch top bunk box spring
105	120	Black smoke out of corridor
156	180	Flames extend to top bunk
190	220	Top bunk fully involved in flames
190	225	Smoke layer drops to bottom of window
210	240	Bunk bed fully involved in flames
275 285		Flashover
278 298		Zero visibility
270	300	Smoke down to 0.30 m (1 ft) above corridor floor
345 345		Window Open
350	350	Flames out of window
420	460	Flames on corridor floor
-	650	Reduction in smoke out of corridor, increase in flames
480	-	Little - no smoke out of corridor
526	740	Limited visibility returned
645	765	Room clear, everything burning
-	900	Flames out of room but not out of corridor
NA	960	Flames no longer extend out of window
1200	-	Bunk bed falls against thermocouple tree
1230 1230		Burnout
1380	-	Fan is turned off
Note: Times were estimated from the video camera views		

Figure 4-22. Exterior View of Doorway to Corridor After the Start of Forced Ventilation (470 s)

Figure 4-23. Exterior View of Window After the Start of Forced Ventilation (380 s)

Figure 4-24. Doorway During Natural Ventilation Experiment (645 s)

Figure 4-25. Window During Natural Ventilation Experiment (470 s)

4.2.1 Heat Release Rate

The peak measured heat release rate was 14 MW for the PPV ventilated fire and approximately 12 MW for the naturally ventilated fire. Peak heat release rates of both fires occurred approximately 40 s after window ventilation with a spike to their respective maximum. The peak of the PPV experiment occurred 5 s after that of the natural experiment. This corresponded to the 5 s period before the PPV fan was started. Comparing the heat release rate between the time of peak and the time where the two curves intersect showed that the PPV created a greater burning rate by approximately 60 % for about 200 s after the fire reached its maximum output. After the heat release rate spiked, the PPV output remained 4 MW above that of the naturally ventilated experiment for 70 s. At the end of those 70 s, the rates converged until 590 s when the naturally ventilated fire had the higher heat release rate. The naturally ventilated fire remained roughly 1 MW above the output of the PPV ventilated fire until the end of the experiment (Figures 4-26 and 4-27). The integral of the heat release rate curve in Figure 4-28 provided the total heat released over the duration of both experiments. The PPV ventilated experiment released 3.7×10^6 kJ and the naturally ventilated experiment released 3.4×10^6 kJ. The fan caused heat to be released quicker in the PPV experiment, but ultimately both experiments released approximately the same amount of heat.

For fires burning in the open under the laboratory hood, the chemical power measured by the oxygen depletion calorimeter was equal to the heat release rate from the fire as a function of time. However, for a fire within a room, the effluent from the enclosure was a mixed average of the upper layer gases, and does not represent the instantaneous heat release rate of the fire. Prior to ventilation there was a delay in the heat release rate measured due to the room configuration and the time needed for combustion products to travel out of corridor doorway. After the window was opened a majority of the burning took place on the exterior of the room which shortened the time between the release of heat by the fire and when it was detected by the oxygen depletion calorimeter.

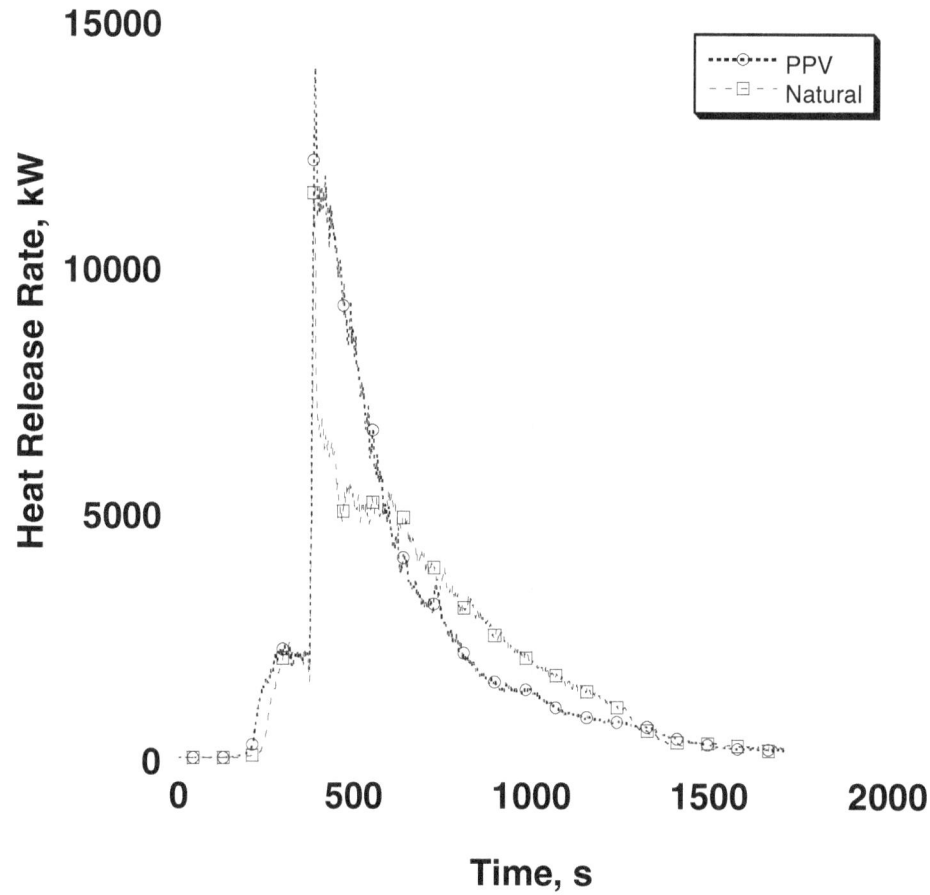

Figure 4-26. Heat Release Rate

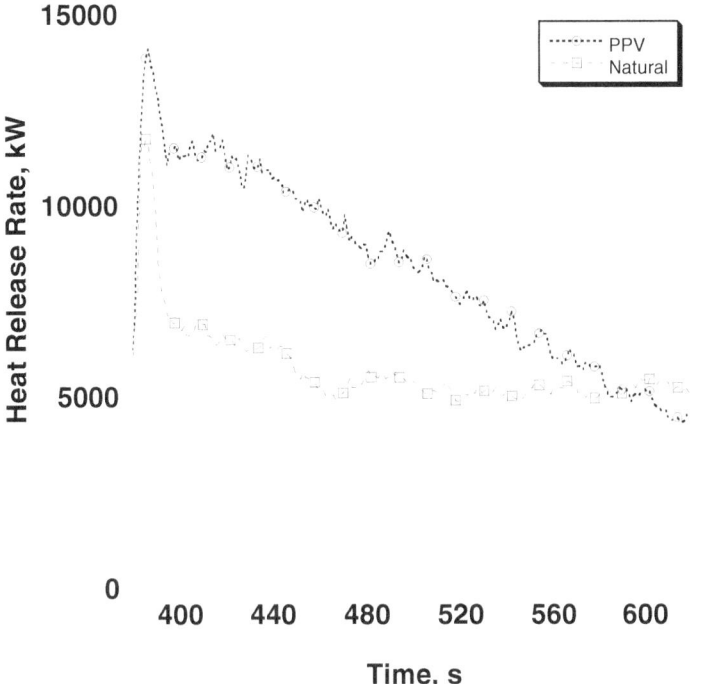

Figure 4-27. Heat Release Rate Detail For 200 s Following Peak Output

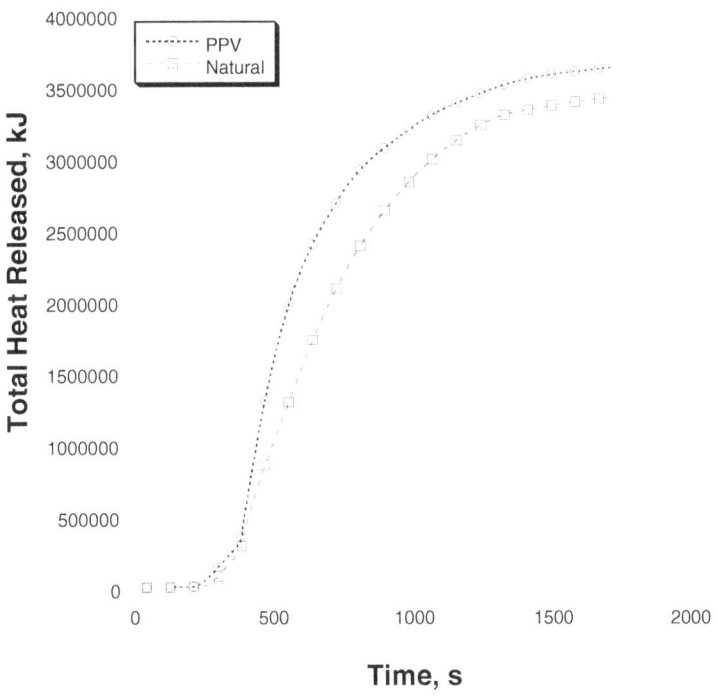

Figure 4-28. Total Heat Released

4.2.2 Room Gas Temperature

The gas temperatures measured in the room were similar for both experiments prior to ventilation as each fire grew to an initial peak of approximately 800 °C (1470 °F) (Figure 4-29). Flashover occurred approximately 270 s after ignition and both fires became ventilation limited. Once the fires were ventilation limited, the upper layer temperatures decreased to 700 °C (1290 °F). When ventilation was started in the experiment with a PPV fan, the upper layer temperature increased temporarily to 800 °C (1470 °F), quickly dropped to 550 °C (1020 °F) and then rapidly increased to the maximum temperature of approximately 980 °C (1800 °F). The maximum temperature was maintained for a short period of time and then the temperatures in the room steadily decreased to 400 °C (750 °F) at a rate of 0.8 °C /s. At 1200 s into the experiment a piece of the burning bunk bed fell onto the thermocouple leads and caused the room gas temperatures shown in Figure 4-29 to be invalid after this point.

The naturally ventilated fire produced a much smoother time evolution of room temperatures. After ventilation, the temperatures rapidly increased to the maximum temperature of 1050 °C (1890 °F). The temperatures remained approximately 1000 °C (1830 °F) for approximately 300 s. Once the temperatures began to decrease the values did so steadily to 500 °C (932 °F) at a rate of 0.8 °C /s. At 1430 s there was a rapid decrease in temperature to 100 °C (210 °F) as the fuel in the room was consumed (Figure 4-30).

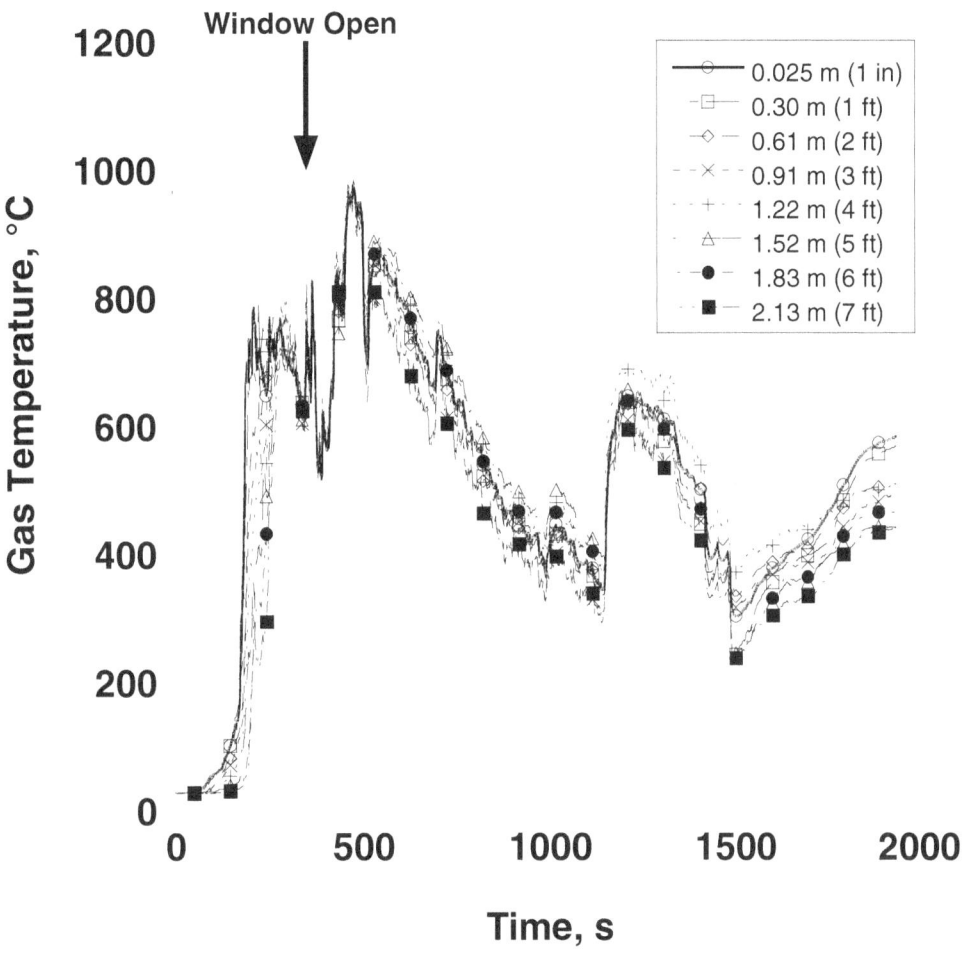

Figure 4-29. PPV Room Temperatures, Distances Measured From Ceiling. Note: A piece of the bunk bed fell on the thermocouple tree at 1200 s, therefore temperatures cannot be compared to the natural ventilation experiment after that point.

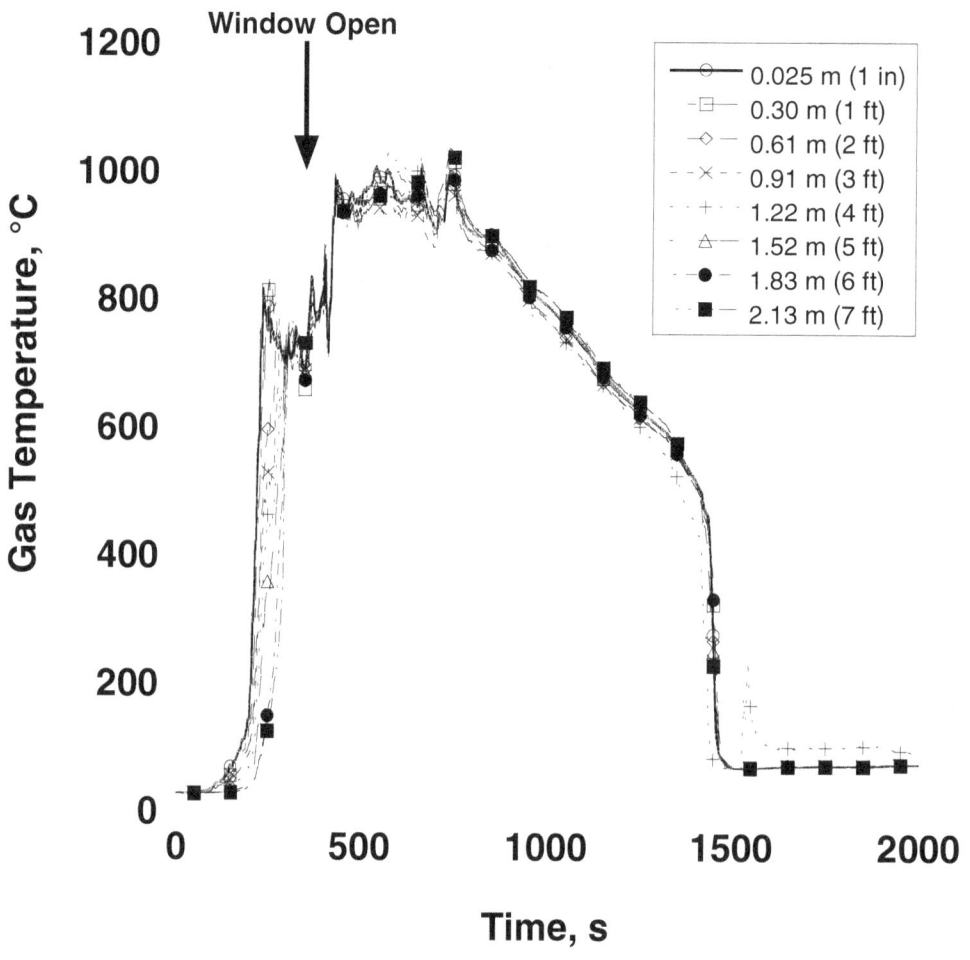

Figure 4-30. Natural Ventilation Room Temperatures

4.2.3 Doorway Gas Temperature

The temperatures recorded in the upper portion of the doorway to the room were comparable to those within the room. For both experiments the temperatures at the top and middle of the doorway were approximately 600 °C (1110 °F) while the lower portion of the doorway remained less than 100 °C (212 °F) prior to window ventilation. This was consistent with the fire drawing ambient air into the room through the lower section of the doorway.

After positive pressure ventilation was initiated, the gas temperatures increased quickly to the peak temperatures of 1000 °C (1830 °F) at the top, 800 °C (1470 °F) in the center and 550 °C (1020 °F) at the bottom of the doorway. Once the fan forced the air into the room, the doorway temperatures began to decline and continued to decrease until the end of the experiment (Figure 4-31). A small increase in temperature occurred at approximately 1380 s which was consistent with the turning off of the fan. The increase was further evidence of the cooling effects of the fan.

The doorway temperatures for the naturally ventilated experiment were higher than those of the PPV experiment for a longer time period. It took approximately 300 s for the maximum temperatures to be reached at the top of the doorway. Both the top and center of the doorway reached a maximum of 1000 °C (1830 °F). The bottom of the doorway briefly peaked at 700 °C (1290 °F) before dropping to 200 °C (390 °F) as the fire continued to burn. Temperatures slowly declined to 100 °C (212 °F) over the 700 s after the peak (Figure 4-32).

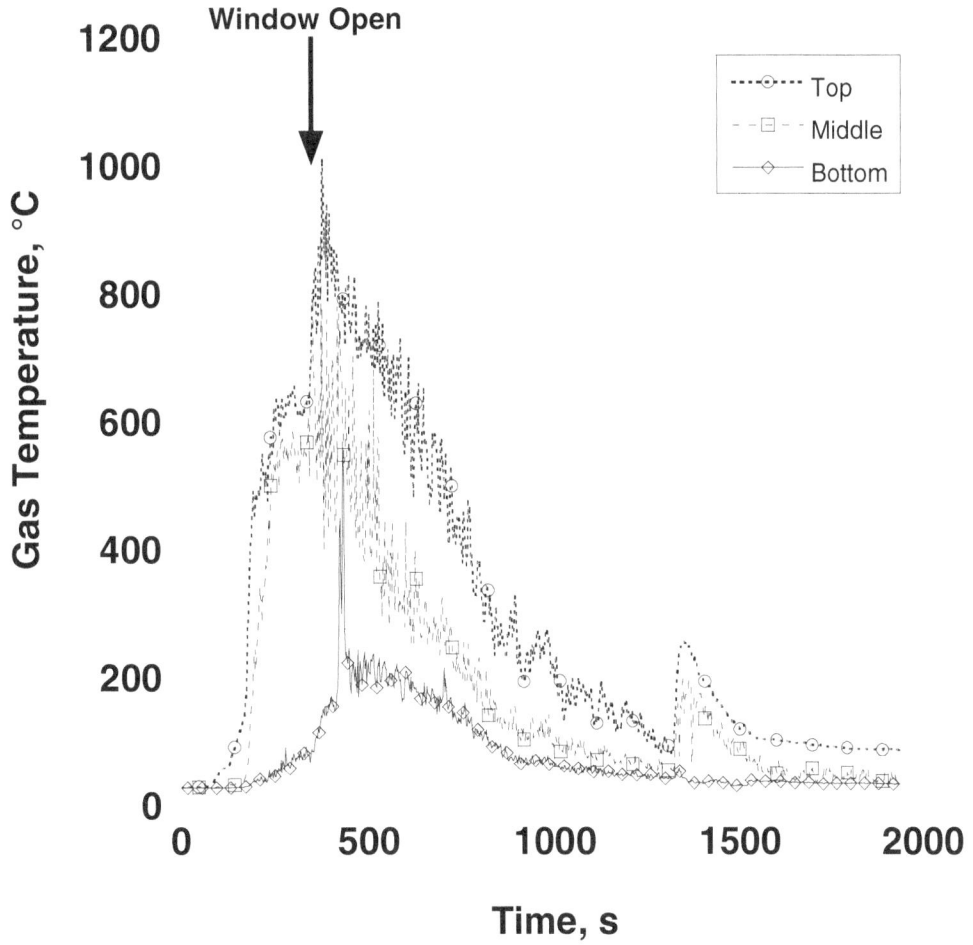

Figure 4-31. PPV Doorway Temperatures

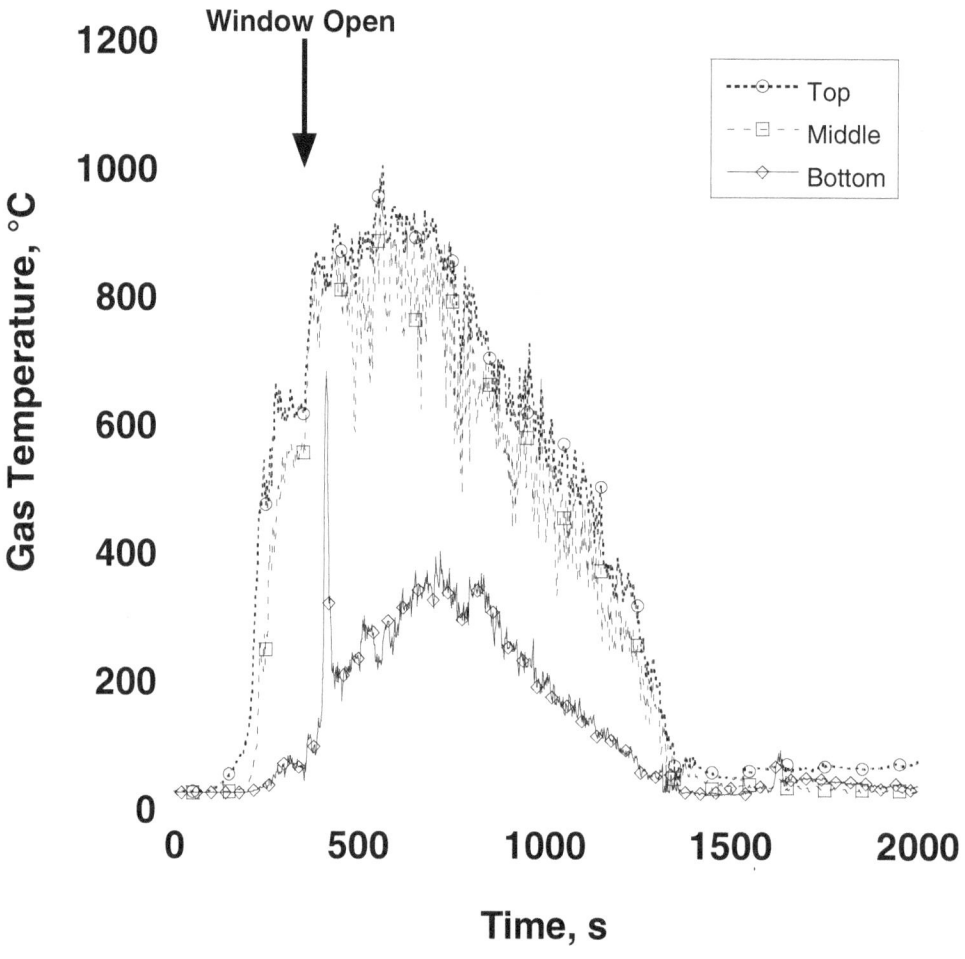

Figure 4-32. Natural Ventilation Doorway Temperatures

4.2.4 Window Gas Temperature

The gas temperatures monitored at the window were significantly different depending on the method of ventilation. The PPV experiment created more uniform gas temperatures in the window due to the unidirectional flow out of the window. Flames and hot gases were observed coming out of the entire cross sectional area of the window. A bidirectional flow pattern existed in the naturally ventilated fire experiment. Flames were seen in the entire window for a short period of time and then air entered the lower third of the window for the remainder of the experiment. The gas temperatures in the PPV experiment were relatively uniform top to bottom between 900 °C (1650 °F) and 1100 °C (2010 °F) while the naturally ventilated experiment had temperatures of 1000 °C (1832 °F) at the top and 600 °C (1110 °F) at the bottom of the window. The PPV experiment required 200 s to reach these temperatures while the naturally ventilated experiment took approximately 400 s (Figures 4-33 and 4-34).

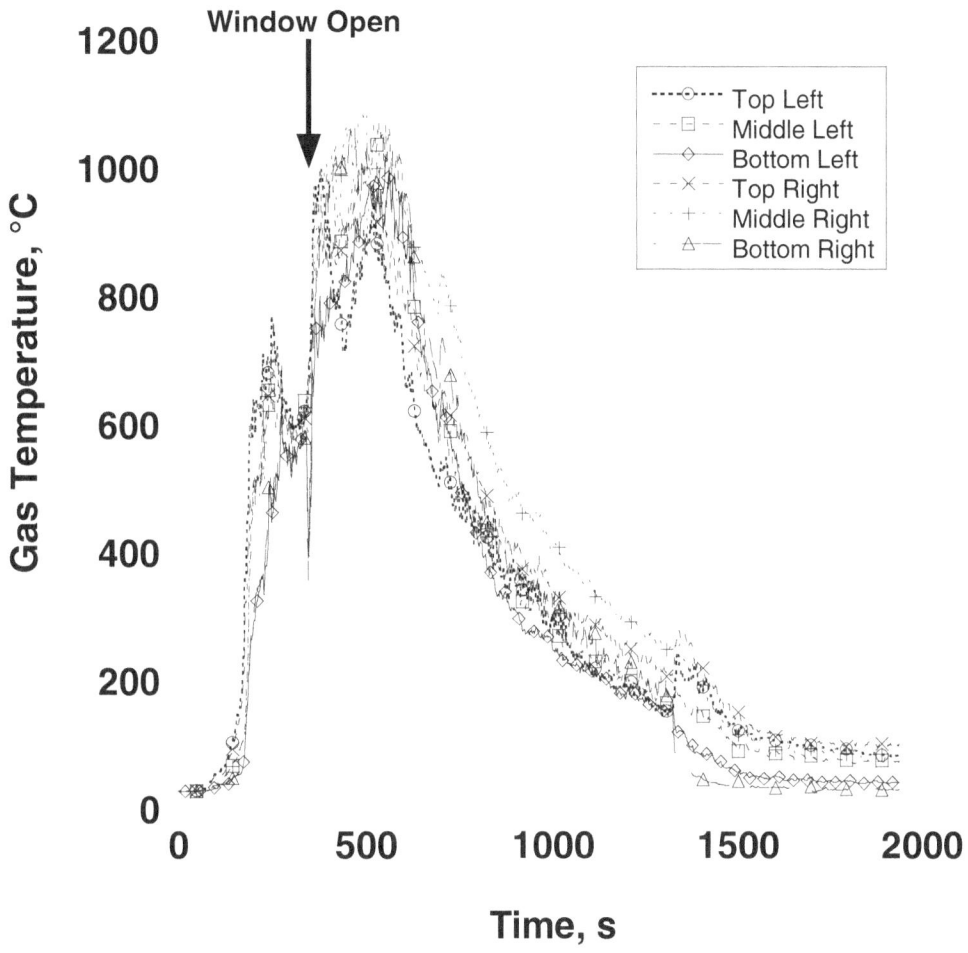

Figure 4-33. PPV Window Temperatures

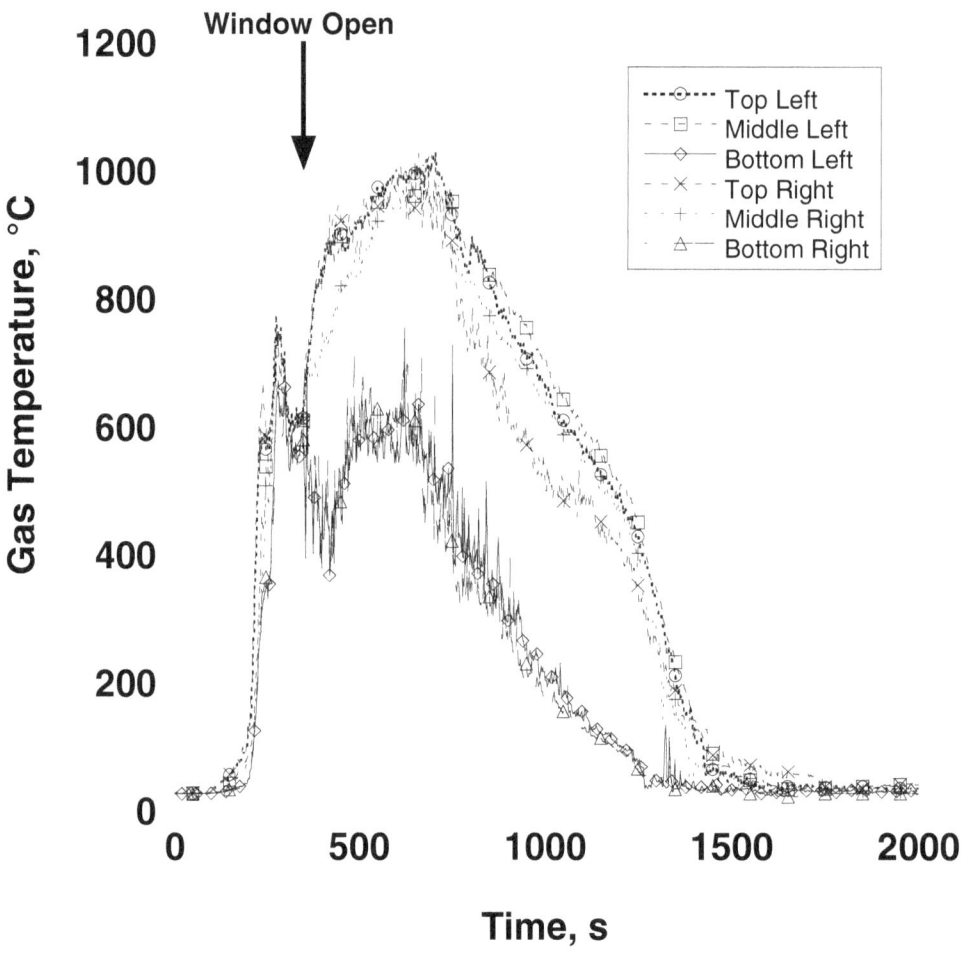

Figure 4-34. Natural Ventilation Window Temperatures

4.2.5 Corridor Gas Temperature

The corridor doorway gas temperatures also showed a significant difference between the two ventilation tactics. Approximately 120 s after the fan was started; the fan reversed the natural tendency for the air to be drawn back into the room. This created a unidirectional flow. After ventilation started in the PPV experiment, the gas temperature reached nearly 700 °C (1290 °F) at the very top of the doorway. Once the fan was turned on, the upper doorway temperatures never increased above 200 °C (390 °F). The bottom half of the doorway remained slightly above ambient temperatures of 25 °C (77 °F).

The naturally ventilated gas temperatures were different due to the flow of combustion gases and flames that ventilated out of the corridor doorway. Gas temperatures in the upper third of the doorway were between 600 °C (1110 °F) and 900 °C (1650 °F) after ventilation (window opened). The mid doorway temperature rose as high as 400 °C (750 °F) while the temperature at the bottom remained approximately 100 °C (212 °F). The temperature trends in the corridor were very similar to those of the room doorway (Figures 4-35 and 4-36).

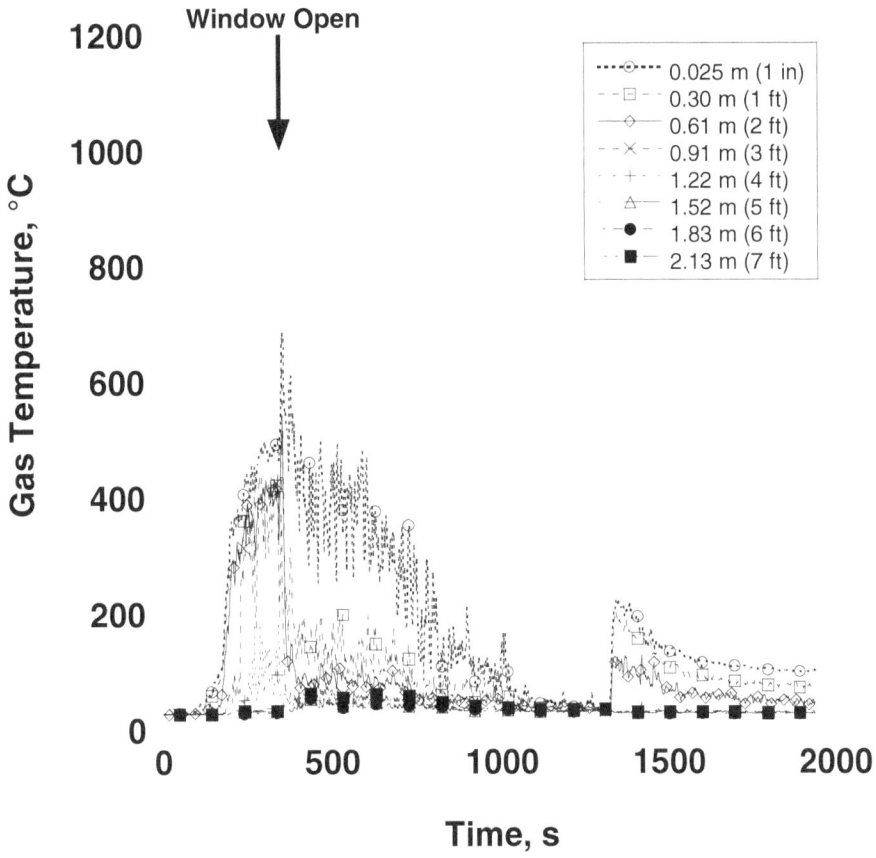

Figure 4-35. PPV Corridor Temperatures

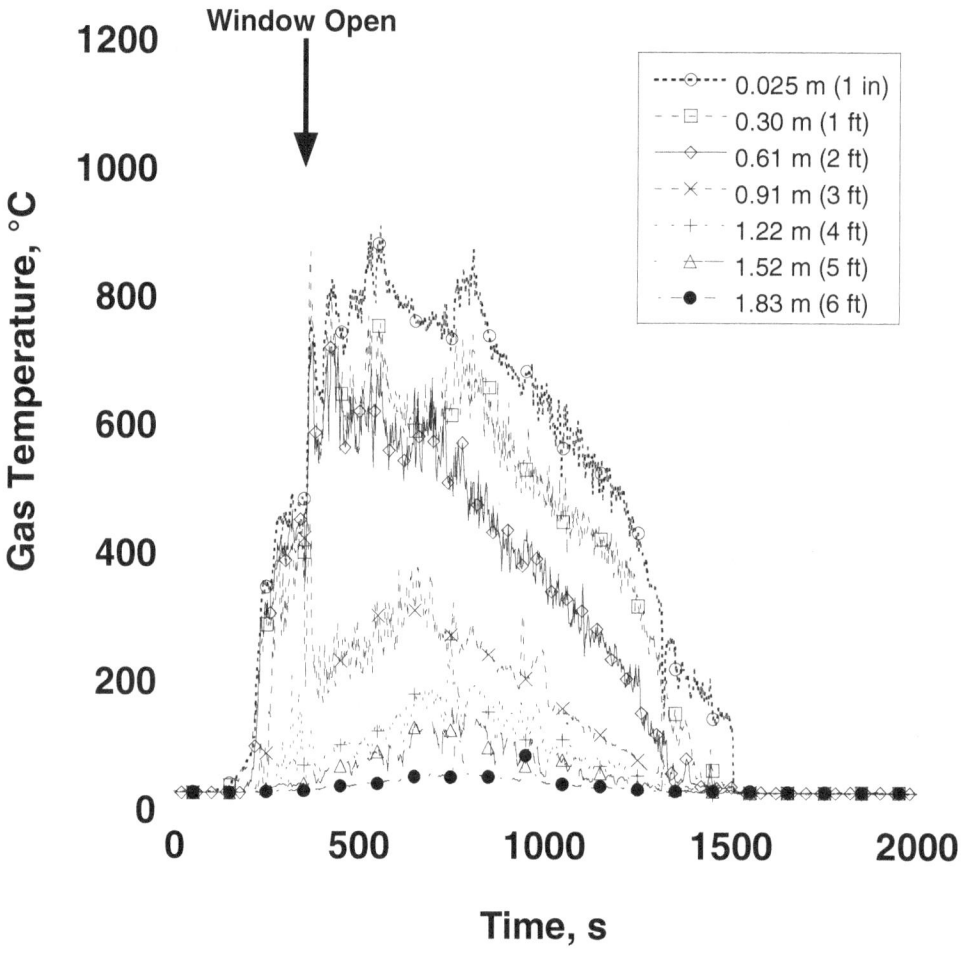

Figure 4-36. Natural Ventilation Corridor Temperatures

4.2.6 Room Differential Pressure

Differential pressure readings were monitored to track the static pressure in the room created by the fire and the impact of the PPV fan on this pressure. The interior pressure readings were referenced to the pressure at the same elevation on the outside of the room. The negative time values represent times prior to ignition. Before ignition in the PPV experiment, the fan created uniform pressures at all three elevations of 21 Pa (0.003 PSI). After ignition, the fire created pressures of 34 Pa (0.005 PSI) at the top probe, 14 Pa (0.002 PSI) at the middle probe and -14 Pa (-0.002 PSI) at the lower probe. Once the window was opened and the fan was turned on, these differential pressures became 62 Pa (0.009 PSI), 41 Pa (0.006 PSI) and 21 Pa (0.003 PSI) respectively. These pressures held constant for a period of time when the fire was at peak and then declined steadily. The naturally ventilated fire created differential pressures of 28 Pa (0.004 PSI) at the top probe, 7 Pa (0.001 PSI) at the middle probe and -14 Pa (-0.002 PSI) at the bottom probe. These differential pressures declined slightly until the fire diminished (Figures 4-37 and 4-39).

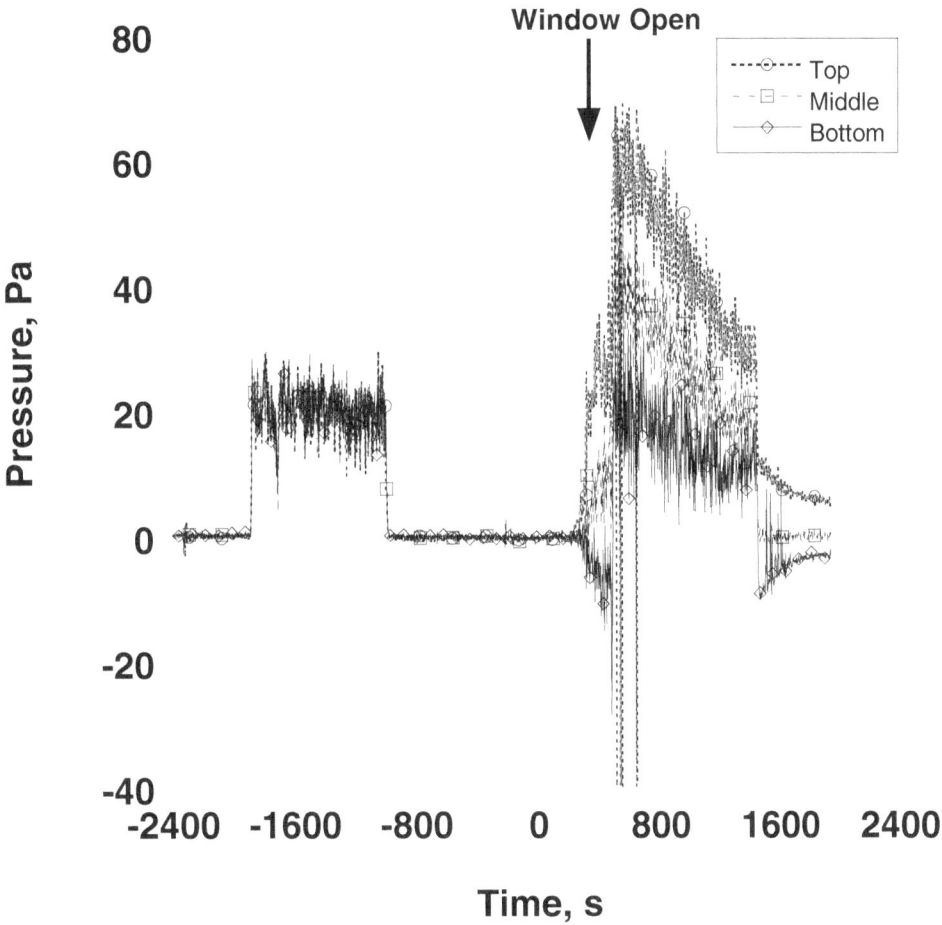

Figure 4-37. PPV Room Differential Pressure

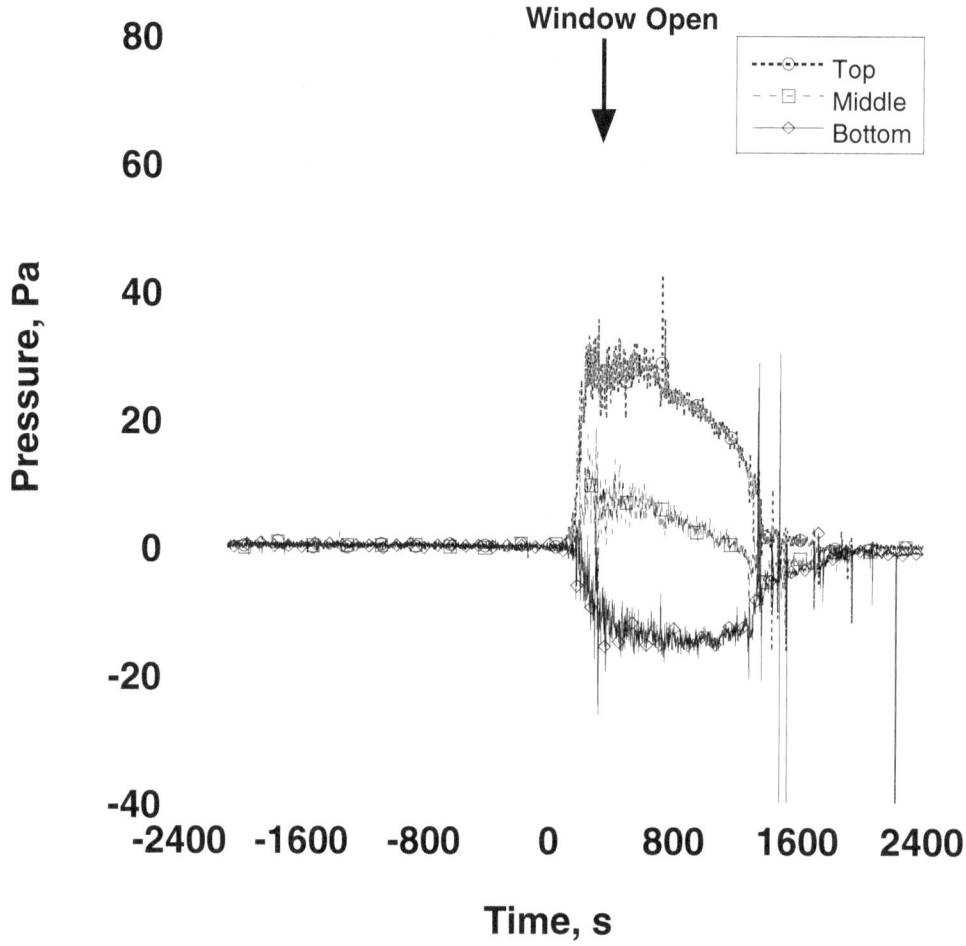

Figure 4-38. Natural Ventilation Room Differential Pressure

4.2.7 Window Gas Velocity

Before the PPV ventilated experiment, an ambient flow experiment (without a fire) of the fan through the room was conducted. The experiment produced an average velocity of 5 m/s (16 ft/s) out of the window. For the PPV experiment, velocities on the order of 5 m/s (16 ft/s) to 20 m/s (66 ft/s) were measured. The highest velocity occurred just after the window was opened and the fan was turned on. With the fire growing, the velocity increased to 20 m/s (66 ft/s) and slowly decreased to the fan velocity of 5 m/s (16 ft/s) as the fire decreased. The naturally ventilated experiment had a bidirectional flow through the window with the highest velocities of 12 m/s (39 ft/s) at the top of the window. The gas velocity in the middle of the window was about 7 m/s (23 ft/s) out of the room while the bottom of the window had a flow into the room of 2 m/s (7 ft/s). It took longer for the maximum velocities to be reached than in the PPV experiment but this was also directly proportional to the growth of the fire (Figures 4-39 and 4-40). In the three minutes following the window being opened, the average gas velocity produced by the PPV experiment was 14 m/s (46 ft/s) while in the naturally ventilated experiment, the average gas velocity was 5.5 m/s (18 ft/s).

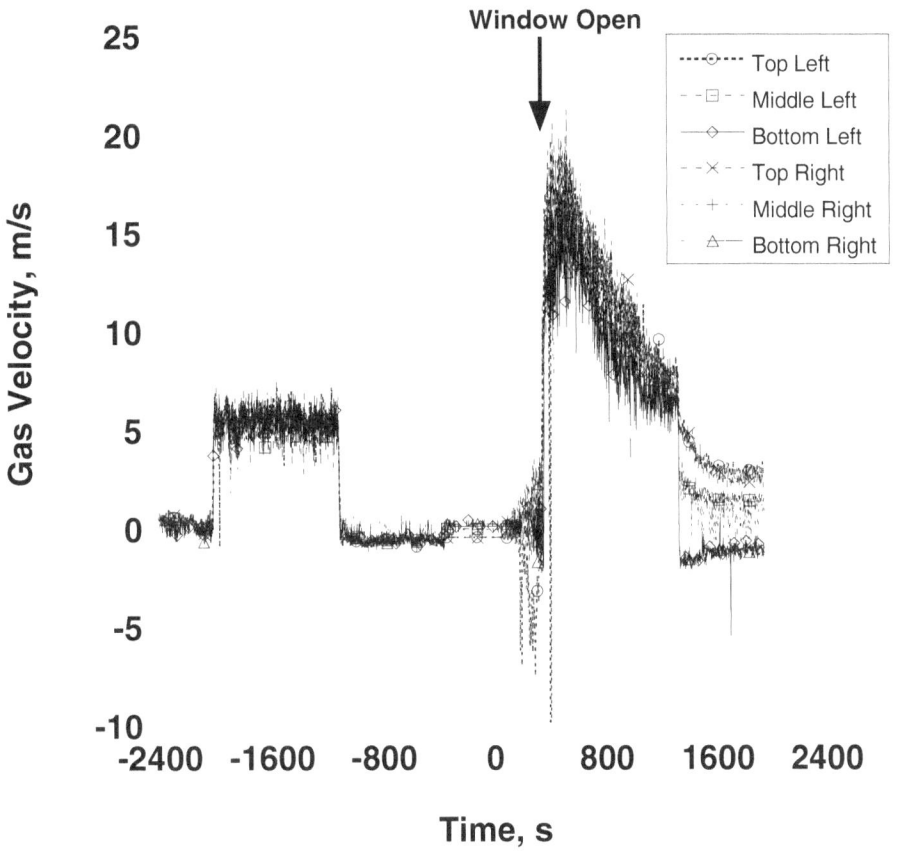

Figure 4-39. PPV Window Velocities

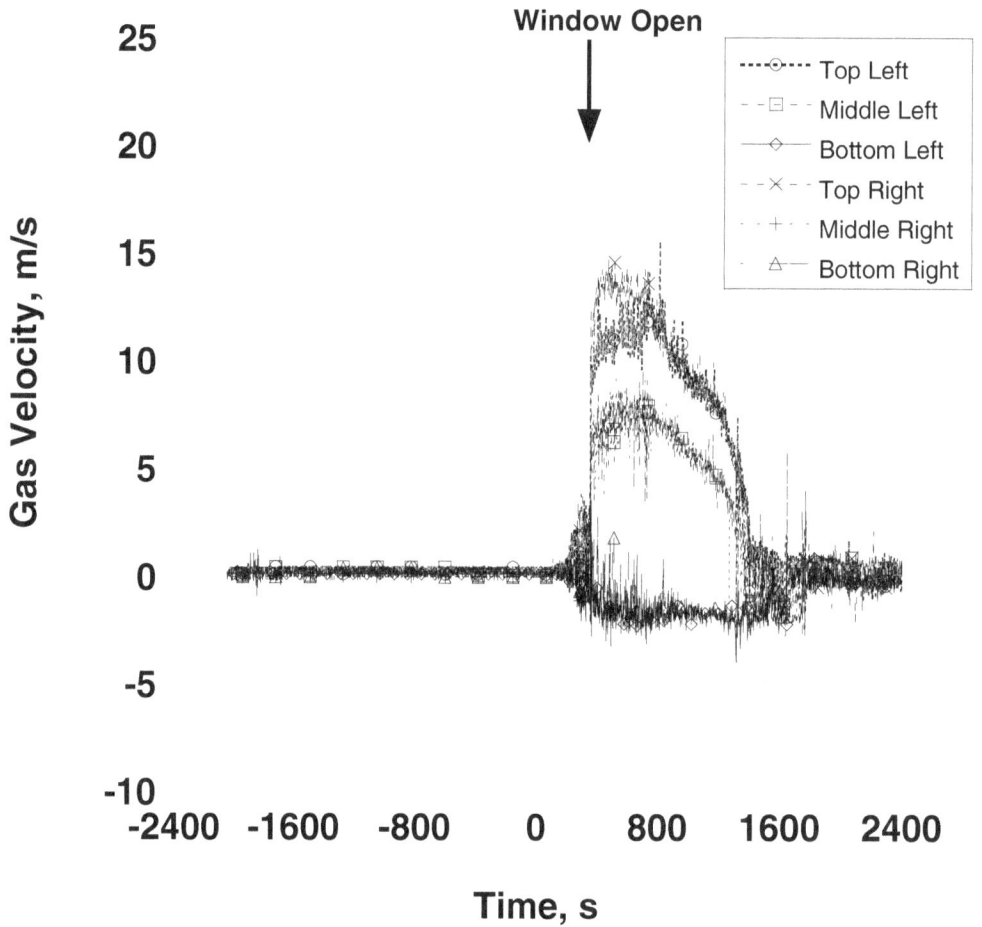

Figure 4-40. Natural Ventilation Window Velocities

4.2.8 Doorway Gas Velocity

The gas velocities into the room through the doorway were lower than those out through the window. The PPV fan alone created average flow velocities of 3 m/s (10 ft/s) to 4 m/s (13 ft/s) as shown in Figure 4-41 for times of -2000 s to -1200 s. The negative time values represent times prior to ignition. Prior to ventilation, there was a 4 m/s (13 ft/s) to 6 m/s (20 ft/s) flow out of the top two-thirds of the doorway and a flow into the room in the bottom one-third of the doorway of 2 m/s (7 ft/s). After the fan was activated, the air flowed into the room via the bottom two-thirds of the doorway and the flow in the upper third of the doorway fluctuated between in and out of the room. Eventually, the fan was able to completely move air into the room over the entire doorway cross section. The naturally ventilated experiment began in the same manner as the PPV experiment with bidirectional flow through the doorway. Once ventilation was started, the gas velocities held rather steady until the fire began to decrease with the velocities decreasing as well.

The flows into the room may be underestimated due to the orientation of the bi-directional probes. The probes were faced into the room and perpendicular to the corridor which may not have measured the full magnitude of the velocity into the room but were accurate for the flows out of the room (Figures 4-41 and 4-42). The flow out of the room had to pass in the direction of the centerline of the probes yielding a more accurate differential pressure as opposed to the flow into the room that may have passed the bi-directional probe at an angle to the centerline of the probe causing lower differential pressure readings, which correspond to under estimation of the gas velocities.

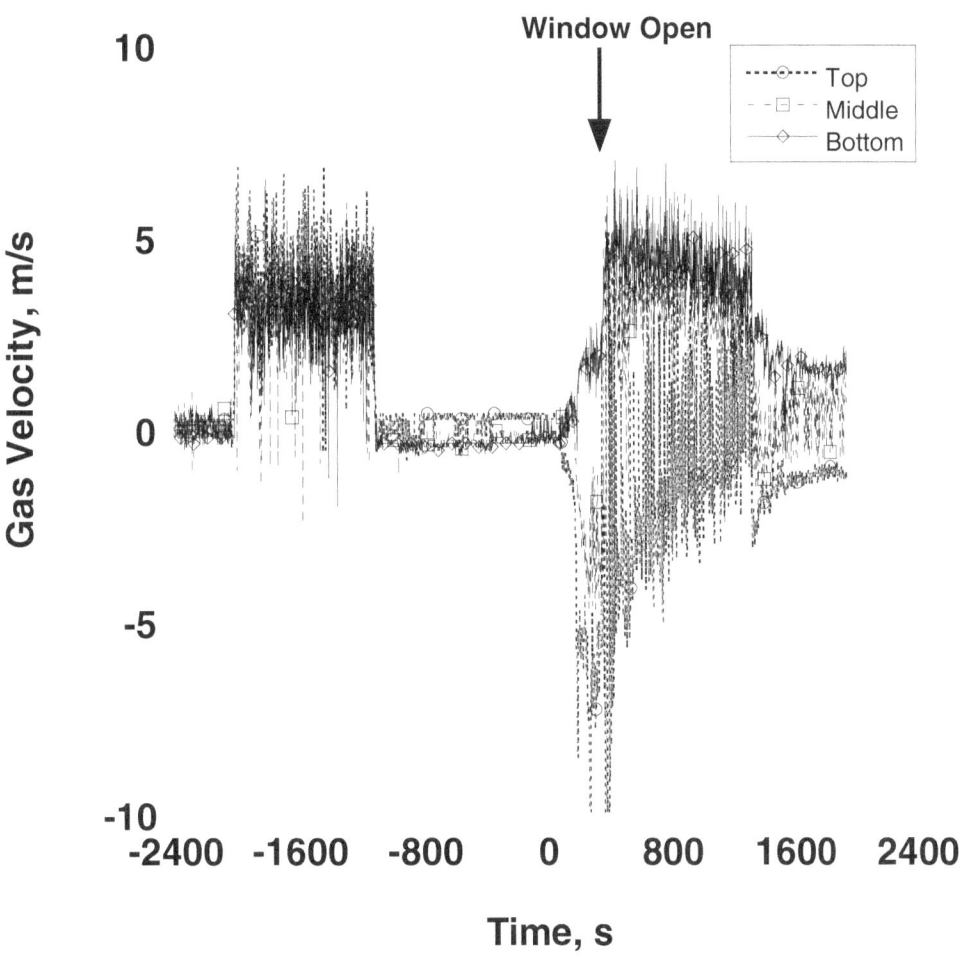

Figure 4-41. PPV Room Doorway Velocities

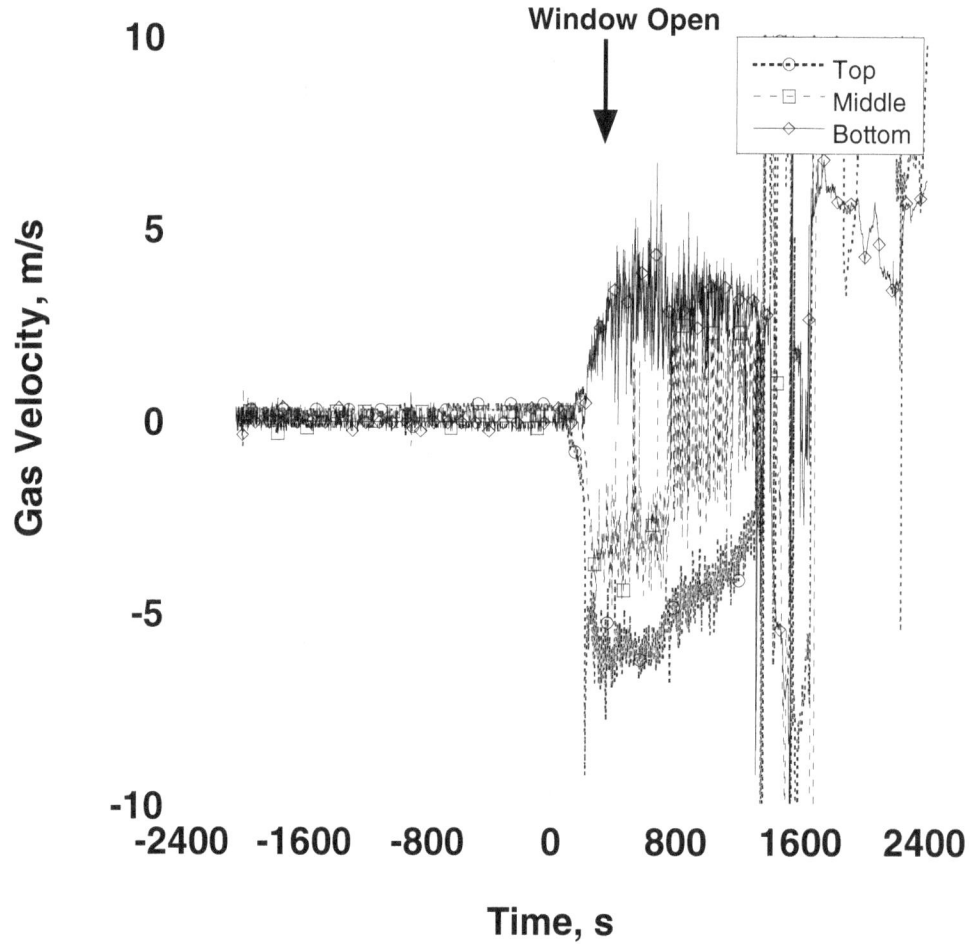

Figure 4-42. Natural Ventilation Room Doorway Velocities

4.3 Computer Simulation

The two experiments documented above were simulated using FDS in an attempt to visualize and accurately quantify the effects of a PPV fan on a post flashover room fire. The outputs of both simulations were then compared to the experimental data collected. Although very complex, FDS was able to replicate the global effects of the PPV fan on the room fire.

4.3.1 Domain

The domain used for the room fire calculation measured 6.0 m (19.7 ft) x 6.2 m (20.3 ft) x 3.0 m (9.8 ft) and was comprised of approximately 380,000 grid cells that measured 0.067 m (2.62 in) on a side (Figure 4-43). The domain was larger than the room in all directions to allow for more accurate pressure differences that would be created by the PPV fan. For the simulation with the PPV fan, the domain characteristics described in chapter 1 were utilized. The grid for the fan and the grid for the room were multi-blocked together for a total of just over 1.6 million grid cells (Figure 4-44).

The grid cell size was selected in order to resolve accurately the furniture in the room and to minimize the difference in grid cell size between the room grid and the fan grid. Numerous variables existed such as fire growth rate and time to flashover that needed to correspond with the experiment without the fan. Therefore, numerous partial simulations were run in order to accurately replicate the variables that took place in the first 345 s. The grid cell size that was selected allowed this fire development time to be simulated in approximately 40 hours. The final simulation was allowed to run for 1500 s until the fire was only smoldering as compared to the experimental data. This run required 15 days of computational time to complete on 3.2 GHz processors. Preliminary runs were completed with grid cell sizes that varied from half to twice the size of the 0.067 m (2.62 in) on a side (Figure 4-45). The runs with larger grid cells developed much slower as compared to the experiment and the smaller grid took too long to complete so grid cell independency can not be claimed for this analysis. The final simulation with the fan was allowed to run for 33 days and produced 588 s of simulation prior to an unexpected power outage. The simulation time was into the decay stage of the fire so it was of sufficient duration to use for comparison to the first portion of the actual PPV experiment.

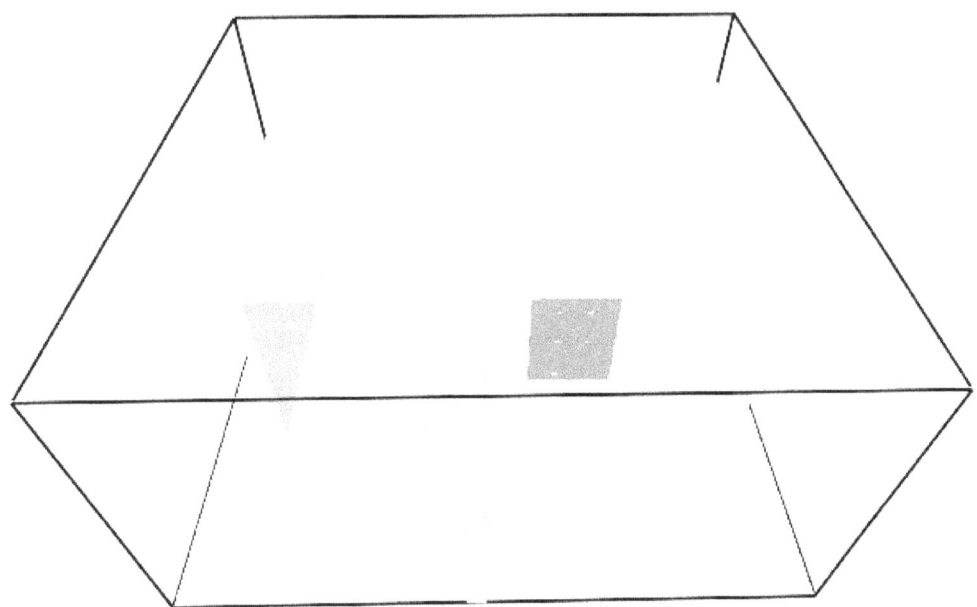

Figure 4-43. FDS Naturally Ventilated Domain

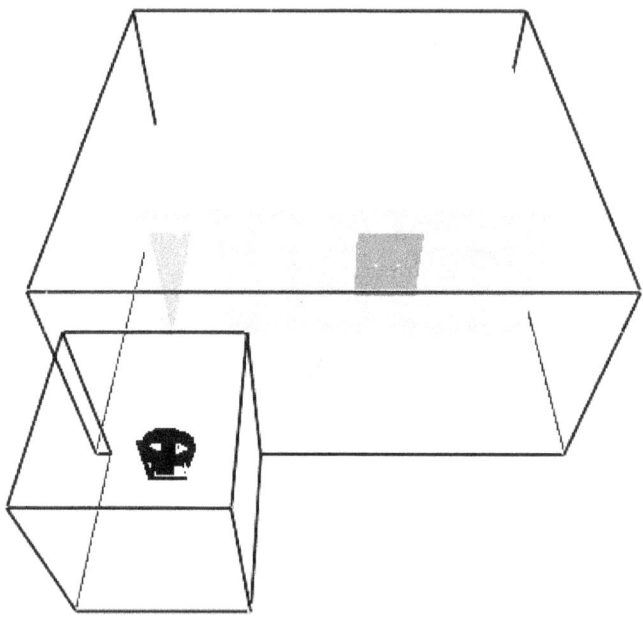

Figure 4-44. FDS PPV Ventilated Domain

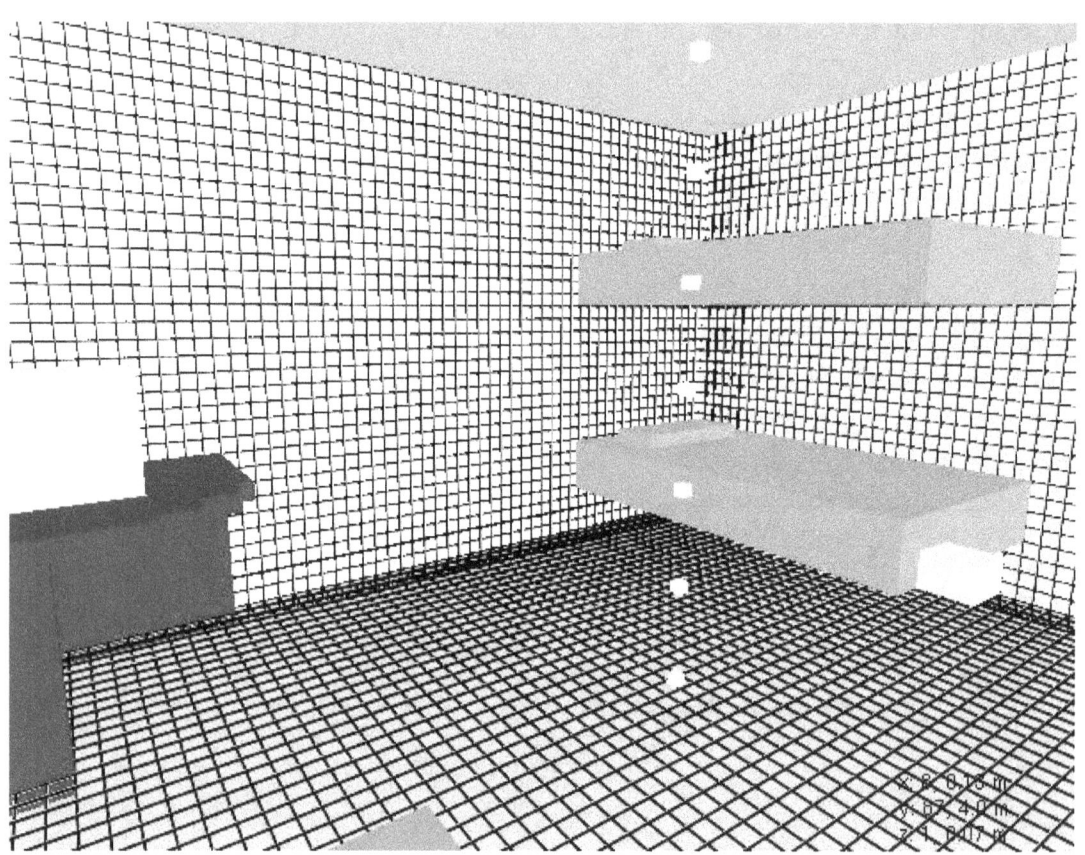

Figure 4-45. Grid Cell Visualization

4.3.2 Geometry

The simulation geometry was input into the model using the dimensions from the experimental floor plan. All of the geometry was prescribed using rectangular obstructions that were forced to conform to the rectilinear grid described in the previous domain description. For this reason the walls and furniture items were not the exact dimensions or in the exact location as they were in the experiment. These slight variations typically had little impact on the calculation and were analyzed to determine the impact of grid cell dependency. Obstructions were also not able to be entered diagonally. For example the computer monitor appeared sideways in the calculation but contained a similar amount of fuel. The front of the desk also appeared stepped for this same reason (Figure 4-46). Subtle differences also existed as the round edges of the chair cushions or mattresses were not able to be exactly captured.

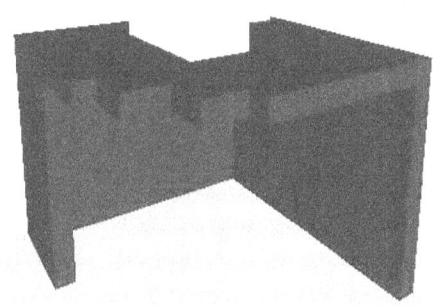

Figure 4-46. Rectangular Geometries

4.3.3 Materials

Each obstruction in the domain was given a set of physical and thermal properties that were used in the calculation. Each wall, ceiling, floor and piece of furniture was defined with properties such as thermal diffusivity, heat of vaporization, density and thickness. These values are shown in Table 8. All of the walls and the ceiling of the room were gypsum board and the floors were covered in carpet, as in the experiments. The primary fuels in the room were upholstery and oak. The chair and mattresses were prescribed as upholstery and the bookcase and desk were both oak. All of the material properties were derived from standard reference literature, not from measurements of the items themselves [25, 26].

Table 8. Room Fire FDS Input Material Properties

Material	Ignition Temperature (°C)	Thickness (m)	Heat Release Rate Per Unit Area (kW/m^2)	Heat of Vaporization (kJ/kg)
Upholstery	280	n/a	n/a	1700
Oak	340	0.02	n/a	4000
Plastic	370	n/a	500	n/a
Carpet	280	n/a	n/a	3000
Gypsum Board	400	0.013	100	n/a

4.3.4 Vents and Ignition Source

All of the external domain boundaries were prescribed as open with the exception of the floor outside the room. This floor was treated as an inert, cold solid boundary. Another small vent measuring 0.067 m (0.2 ft) by 0.13 m (0.4 ft) used for ignition was located at the corner of the bottom mattress (Figure 4-47). The vent had a constant heat release rate of 25.0 kW. This vent produced heat for the duration of the simulation. The location of the vent corresponded with the location of the electric match in the experiments.

Initially the only openings within the domain were the door to the corridor and the door to the room. The doorways both measured 0.9 m (3.0 ft) wide and 2.0 m (6.6 ft) high. At 345 s the room window was opened as it was in the experiments. The domain extended outside the door and window to allow for the combustion products to flow out of the room. When the fan was added for the PPV simulation the boundaries to its domain were also open and the vents for the fan were located as described in chapter 1.

Figure 4-47. Location of Ignition Source

4.3.5 Output Files

Output files were prescribed to best replicate the measurements taken during the experiments. Vertical and horizontal temperature and velocity slices were placed through the center of the room, corridor, each doorway and the window. Velocity and temperature measurement points were also prescribed at the same locations as in the experiment layout using FDS's thermocouple input. Heat release rate was continually recorded in the domain as well as the flame isosurface locations. A three dimensional smoke was also used for comparisons of smoke movement and visibility.

4.4 Simulation Results

4.4.1 Naturally Ventilated Simulation

The results of the naturally ventilated simulation were compared with the video record of the experiment and the measurements of gas temperature, gas velocity and heat release rate. Visual comparisons of the experiment and simulation are shown in Figures 4-48 through 4-59. Quantitative comparisons between the experimental data and the model predictions are given in Figures 4-60 through 4-66.

4.4.1.1 Visual Comparisons

Figures 4-48 through 4-59 are composed of pairs of images. The still frames captured from experimental video tape appear on the left. The frames were not all taken from the same camera view. The camera view is included in the figure caption because some of the views are obstructed by smoke and or flames. The images on the right were rendered in Smokeview. Both images represent the same time after ignition. The images shown are at times 0, 90, 230, 275, 300, 330, 360, 430, 540 and 720 s after ignition.

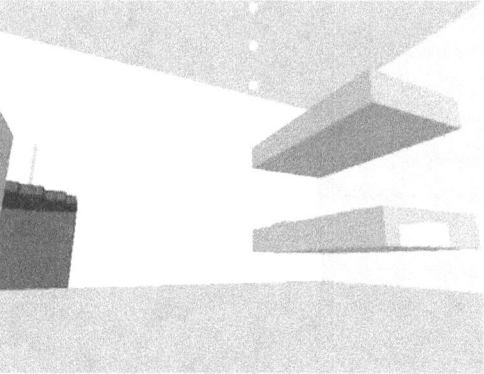

Figure 4-48. Bunkbed View at Time of Ignition (0 s)

Figure 4-49. Doorway and Window View at Time of Ignition (0 s)

Figures 4-48 and 4-49 show the two interior camera views just before ignition. Figure 4-48 shows the gypsum board walls and ceiling, carpeted floor, bunk bed covered with linens on the right, and the desk and computer monitor on the left. The thermocouple tree in the center can also be seen. The image from the simulation, on the right, the comparable finishes and fuel items can be seen. The yellow cubes represent the locations of the thermocouples.

Figure 4-49 shows the opposite corner of the room comprised of the wall with the room door and the wall with the window. The chair was located in the corner with some of the bunk bed and desk visible on the edges of the image. The room thermocouple tree was also located in this view. The image from the simulation demonstrates the same items. The white stripes that appear in the experimental video images are the locations of the gypsum board joints that have been spackled.

Figure 4-50. Fire Starting on Corner of Mattress (90 s)

Figure 4-50 compares the fire development at 90 s after ignition. A flame was located on the corner of the bottom mattress. In the case of FDS, the yellow block was the simulated ignition source. The area that appeared to be involved with flames was based on the stoichiometric mixture fraction, where there was the ideal mixture of fuel and oxygen for a robust flame to exist. The heat release rate per unit volume represented by the simulated flames was 428.6 kW/m^3. The simulation was not able to account for the drop down of flaming material that takes place in the experiment.

 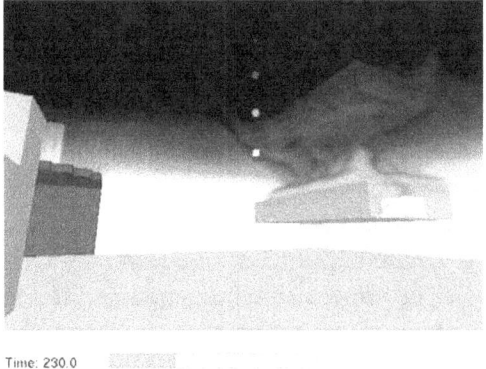

Figure 4-51. Flames Involving Bunkbed (230 s)

The experimental video frame to the left in Figure 4-51 shows fire development 230 s after ignition. The flames have spread across the bottom mattress and have involved the top box spring and mattress. The amount of drop down material had increased due to the burning, melting and falling of the bedding materials and foam mattress. A smoke layer had developed and had begun to descend. To the right the FDS simulation was behaving similarly with slightly less flaming materials. The smoke in the simulation appeared to compare well qualitatively with the experimental video frame.

Figure 4-52. Onset of Flashover (275 s)

Figure 4-53. Visibility Lost in Bunk Bed View (300 s)

At 275 s after ignition the room has gone to flashover as can be seen in Figure 4-52. In the experiment, the flames were burning the carpet and the bunk bed to the left. The smoke had thickened and the layer lowered to the floor as the fire transitioned to an oxygen deficient state. The FDS simulation on the right has also gone to flashover and showed similar flaming and smoke yield as the experimental video frame.

Figure 4-53 has images captured at 300 s after ignition. In both the experiment and the FDS simulation, visibility had decreased to almost zero due to smoke filling. The experimental video was completely black while some of the chair was seen in the left portion of the FDS simulation frame. Even though the door to the room is open to the corridor the camera view suggested that there was little if any fresh air being drawn into the room.

Figure 4-54. Combustion Products Flow From Corridor Doorway (330 s)

While visibility was lost in the room, the external door view in Figure 4-54, at 330 s, showed thick black smoke pouring from the top two thirds of the corridor doorway. There did not appear to be flames coming from the corridor door in the experiment but the FDS simulation was representing flames in this region. This could be due to the difficulty that FDS had when simulating oxygen limited scenarios. By representing the flames in that region FDS predicted that there was enough fuel and oxygen to have combustion but the temperature may not have been high enough to actually have combustion.

Figure 4-55. Flames Extend From Corridor Doorway Once Window is Opened (360 s)

The images in Figure 4-55 show the corridor doorway view after the window was opened. Flames began to extend out of the corridor doorway and the amount of smoke decreased as the fire transitions to a fuel limited stage with the additional oxygen provided by the open window. The flames were also seen over the separation wall coming from the open window. The FDS simulation also showed flames coming from both openings and a decrease in smoke production.

Figure 4-56. Flames From Window (360 s)

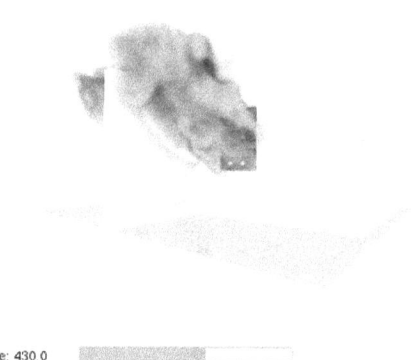

Figure 4-57. Flames Continue From Window (430 s)

Figure 4-56 has images taken from the window side of the separation wall at 360 s after ignition. This was 15 s after the window was opened. Both images showed flames filling the entire cross section of the window and extending above the height of the room. (The square piece of plywood that can be seen outside the window was used in support of another project and was far enough away from the window to have no effect on the fire flows.)

The same window view is shown in Figure 4-57. Seventy seconds after the previous Figure, there were still flames coming from the entire cross section of the window in both the experimental video frame and the FDS simulation.

77

Figure 4-58. Fire in Decay Stage (540 s)

Figure 4-59. Room Continues to Burn (720 s)

The fire began to decay in both the experiment and the simulation as shown in Figure 4-58. The experiment had thick black smoke and some flames coming from the doorway. Flames were also seen coming from the window in the experiment video frame in Figure 4-59. The FDS simulation under predicted both the amount of smoke and flames present in both figures. The smoke was thinner and the flames pulled into the room from the corridor. Flames were still coming from the window in the FDS simulation but not at the same magnitude and intensity of that in the experiment. The FDS simulation appeared to run out of fuel before the actual experiment.

4.4.1.2 Numerical Comparisons

In this section, values of heat release rate, gas temperatures and gas velocities generated by the naturally ventilated FDS simulation are compared to measurements from the full-scale experiments described in Section 4.2.

Figure 4-60 compares the measured heat release rate from the experiment and the heat release rate predicted by FDS. The experimental peak was 12 MW shortly after the window was opened and the FDS value was approximately 11 MW. The FDS peak occurred within 30 s of the experimental peak. The FDS heat release rate increased more quickly than the experimental values but this can be expected because the fire is in an enclosure. In FDS the heat release rate was measured immediately throughout the domain while the actual experiment did not record any measurable heat release rate until the combustion products left the room and traveled through the calorimetry hood. Both cases showed a large increase once the window was opened and a quick decline from the maximum values once post flashover conditions occurred. The FDS simulation held its peak value for a longer duration than the actual experiment and declined at a faster rate but both cases seemed to burn for a similar duration according to the total amount of heat released (Figure 4-61).

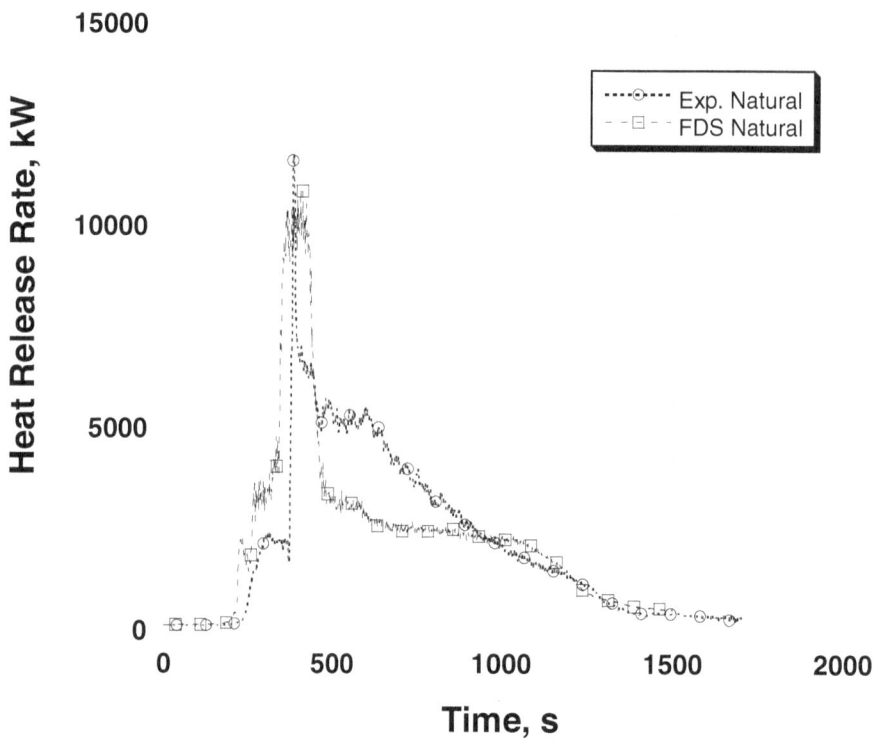

Figure 4-60. FDS and Experimental Naturally Ventilated Heat Release Rate

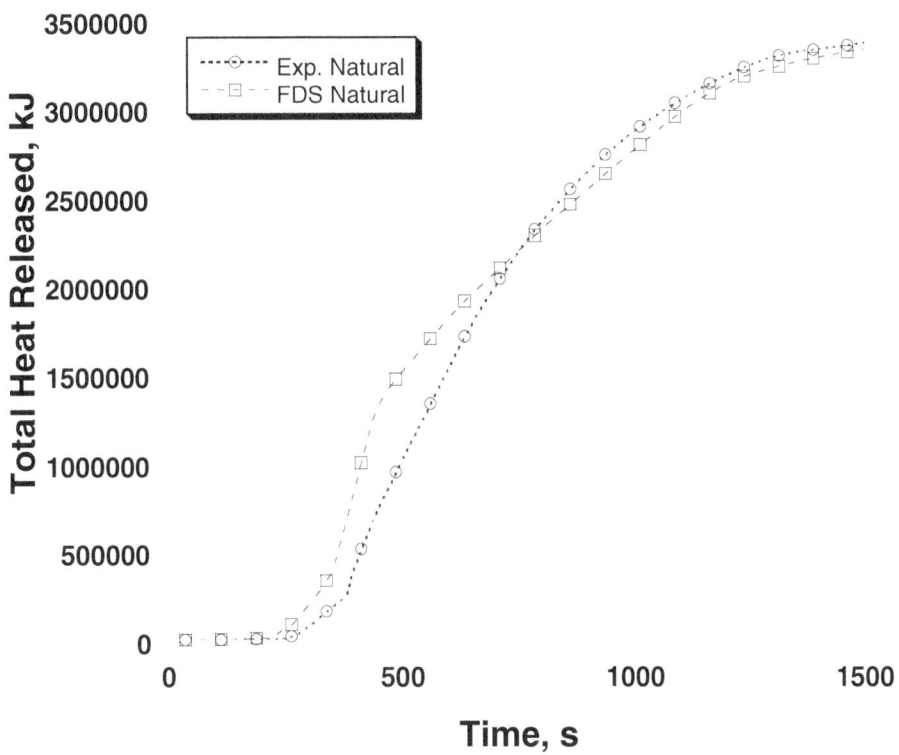

Figure 4-61. FDS and Experimental Naturally Ventilated Total Heat Released

Gas temperatures, both measured and predicted, increased rapidly both before and after ventilation (Figure 4-30 on page 55). In both the experiment and the FDS simulation, gas temperatures rose to approximately 800 °C (1470 °F) prior to ventilation and began to decrease as the room became oxygen limited. Immediately after the window was opened, the room temperatures increased to a peak of 1000 °C (1830 °F) in the experiment and 1100 °C (2000 °F) in the FDS simulation. There was a difference after ventilation in the lower level temperatures. The experiment produced more uniform temperatures throughout the room, while the FDS simulation yielded much lower temperatures lower in the room. Part of this discrepancy can be attributed to the strong radiative heating of the lower thermocouples by the hot upper layer in the experiments which is not included in FDS since the model predicts the gas temperature and not the thermocouple temperature. Previous work done at NIST has demonstrated that lower layer thermocouple readings can be as much as 225 °C (440 °F) below the actual gas temperature [13,14]. This can account for a portion of the difference between lower layer temperatures in figure 4-30 and 4-62, but not all, suggesting that FDS is not capturing all of the details of the flow in the experiment.

After the simulation reached its peak temperature and transitioned to a free burn stage the temperature at the ceiling (0.025 m (0.08 ft)) was slightly over-predicted, the temperatures at 0.3 m (1 ft) and 0.61 m (2 ft) from the ceiling matched very well with the experimental room temperatures. At 0.91 m (3 ft) below the ceiling FDS slightly under-predicted the experimental temperature. The three lowest temperature measurements were under-predicted by 400 °C (750 °F).

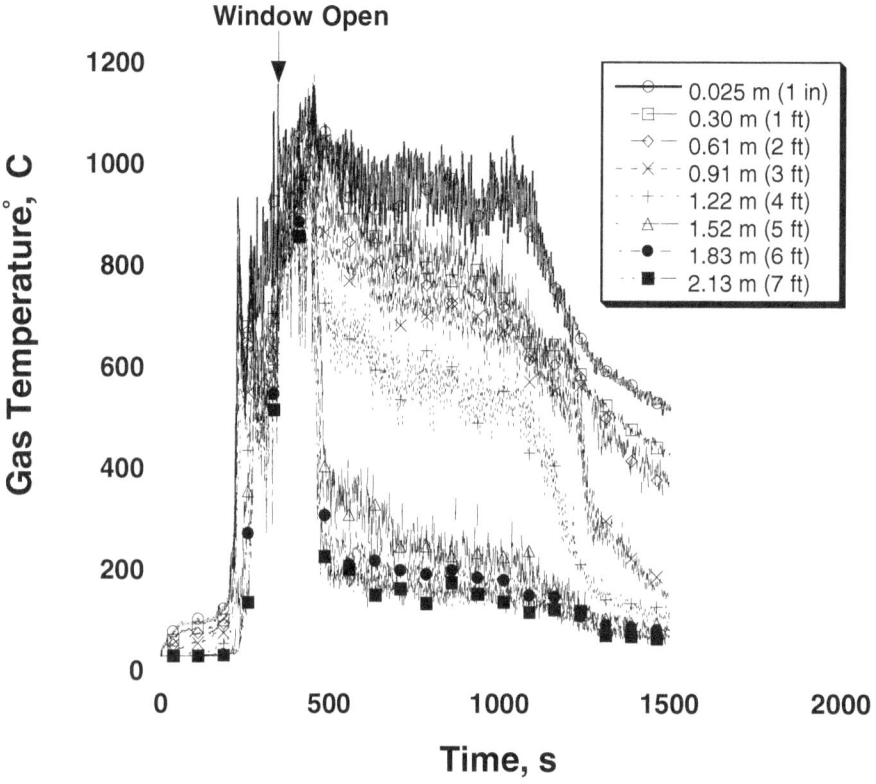

Figure 4-62. FDS Naturally Ventilated Room Temperatures

Figure 4-63 displays the FDS simulation fire room door temperatures. In both the simulation and the experiment, the top and middle measurement points increased to approximately 800 °C (1470 °F) prior to ventilation. After ventilation the experimental fire room door temperatures peaked at 1000 °C (1830 °F) (Figure 4-32 on page 57) while the simulation predicted temperature peaks at 1100 °C (2000 °F). Similar to the lower level room temperatures, the simulation middle temperature in the doorway was significantly lower than in the experiment. The bottom measurement point in both the experiment and the simulation remained low with a peak at the time of ventilation. The FDS simulation also under-predicted this temperature as compared to the experiment.

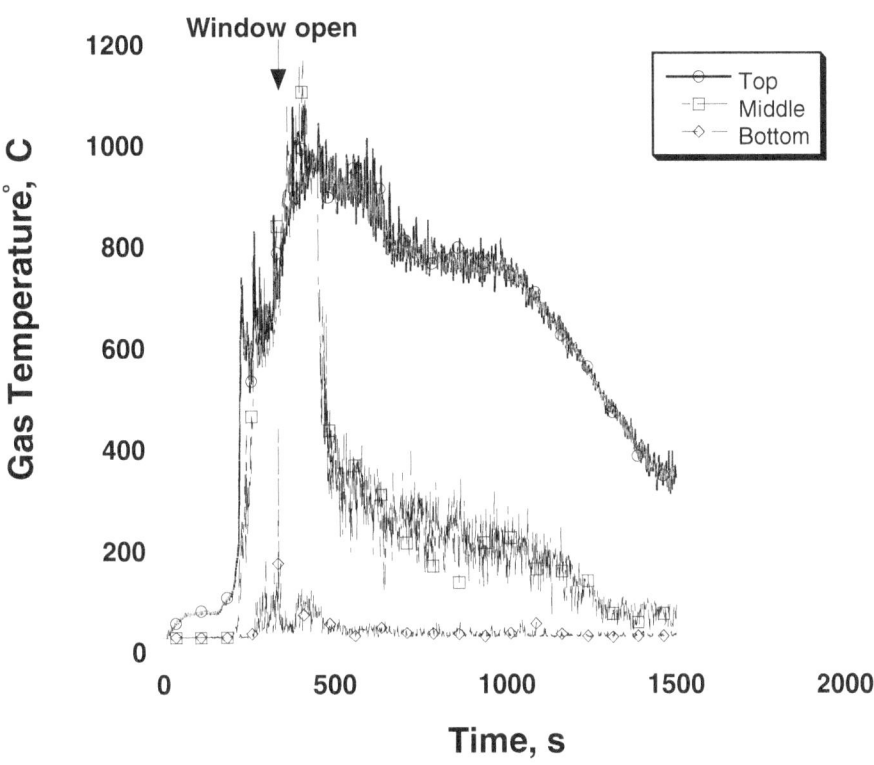

Figure 4-63. FDS Naturally Ventilated Room Doorway Temperatures

Experimental gas temperatures in the upper third of the corridor doorway were between 600 °C (1110 °F) and 900 °C (1650 °F) after ventilation (window opened) (Figure 4-35 on page 60). Simulation gas temperatures at the top of the doorway were higher, peaking at 1300 °C (2370 °F) (Figure 4-64). After the peak the top three temperature measurements in both cases progressed similarly. The experimental mid-doorway temperature rose as high as 400 °C (750 °F) while the bottom remained approximately 100 °C (212 °F). The simulation predicted temperatures at the low level were under-predicted just as they were in the room doorway and in the center of the room. The temperature trends in the corridor were very similar to those of the room doorway.

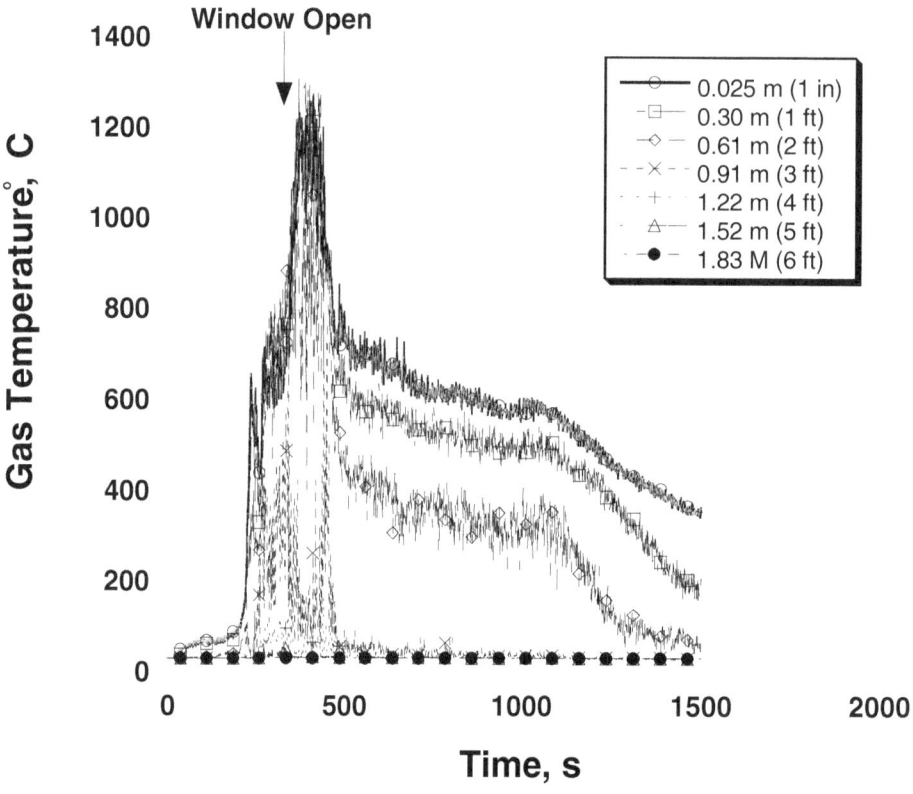

Figure 4-64. FDS Naturally Ventilated Corridor Doorway Temperatures

The naturally ventilated experiment had a bidirectional flow through the window with the highest velocities of 12 m/s (39 ft/s) at the top of the window (Figure 4-40 on page 65). The simulation window velocity peaked at 10 m/s (33 ft/s) at the top of the window (Figure 4-65). Velocities at the top of the window decreased linearly to 5 m/s (16 ft/s) at 1000 s in both the experiment and the simulation. The experimental gas velocity in the middle of the window was about 7 m/s (23 ft/s) out of the room while the bottom of the window had flow into the room of 2 m/s (7 ft/s). The lower window velocities in the simulation were lower in magnitude but were in the same direction. The simulation's mid-window velocities were approximately 5 m/s (16 ft/s) and the lower window velocities were approximately 1 m/s (3 ft/s) out into the window. In both the experiment and the simulation there was a brief time just after the window was opened where the lower window flow is out of the room. Shortly after this, the flow switched into the room as the fire drew air into the room.

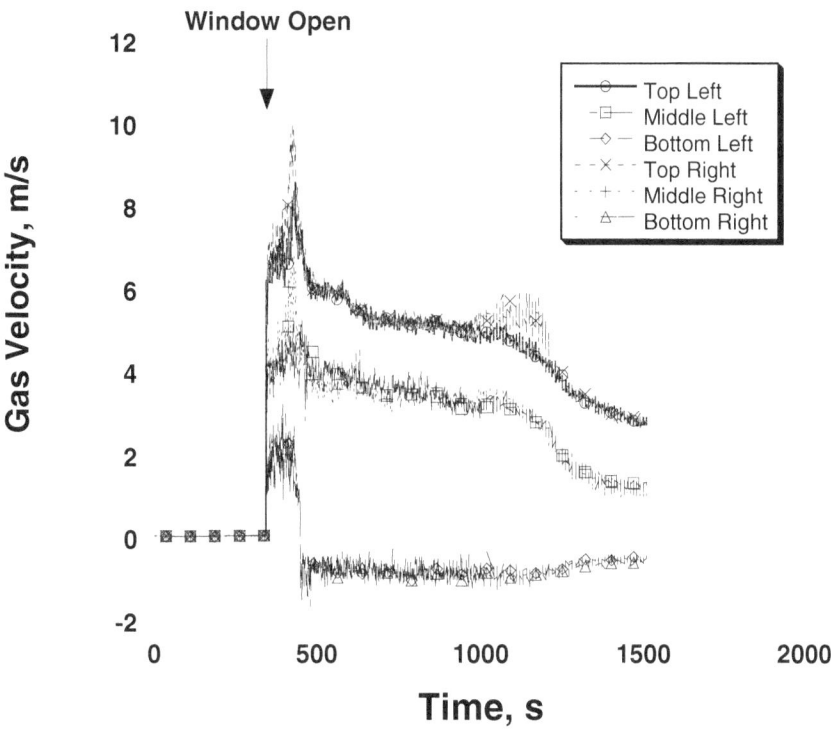

Figure 4-65. FDS Naturally Ventilated Window Velocities

Experimental and simulation fire room doorway velocities are shown in Figures 4-42 (on page 67) and 4-66. Both cases had flow out of the top of the doorway and flow into the bottom two-thirds of the doorway. The experimental flow out of the top of the door fluctuated between 3 m/s (10 ft/s) and 4 m/s (13 ft/s) and the simulation flow ranged between 2 m/s (7 ft/s) and 3 m/s (10 ft/s). Flows into the room in the lower portions of the door were between 3 m/s (10 ft/s) and 6 m/s (20 ft/s) in both the experiment and the simulation.

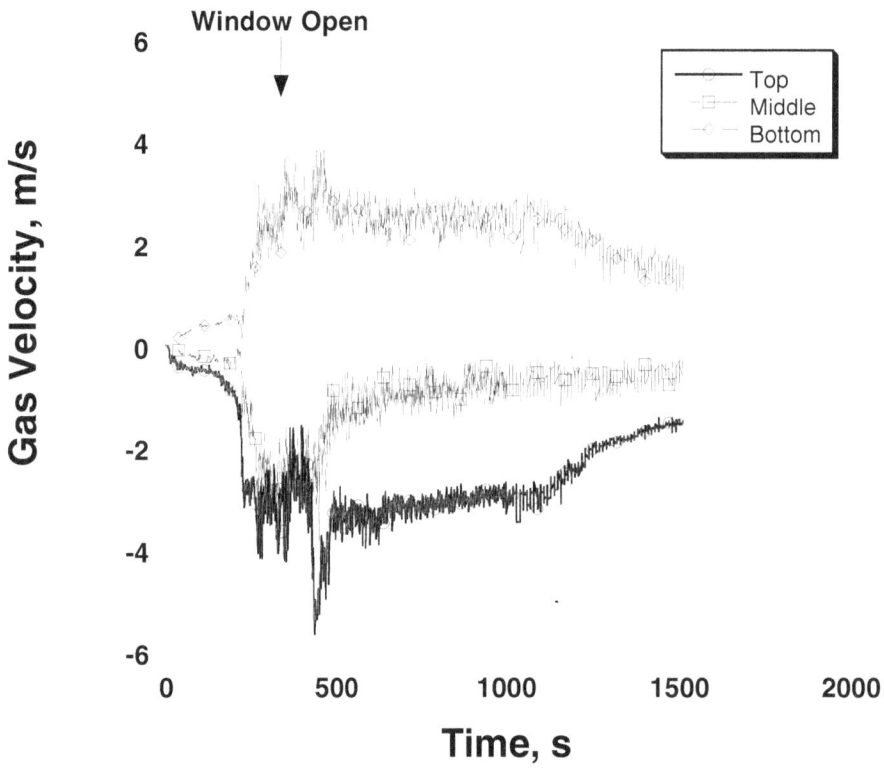

Figure 4-66. FDS Naturally Ventilated Room Doorway Velocities

4.4.2 Positive Pressure Ventilated Simulation

The results of the positive pressure ventilated simulation are compared with the video record of the experiment and the measurements of gas temperature, gas velocity and heat release rate. Visual comparisons of the experiment and simulation are shown in Figures 4-67 through 4-79. Quantitative comparisons between the experimental data and the model predictions are given in Figures 4-80 through 4-86.

4.4.2.1 Visual Comparisons

Figures 4-67 through 4-79 are composed of pairs of images. The still frames captured from experimental video tape appear on the left. The frames were not all taken from the same camera view. The camera view is included in the figure caption because some of the views are obstructed by smoke and or flames. The images on the right were rendered in Smokeview. Both images represent the same time after ignition. The images shown are at times 0, 60, 170, 240, 300, 360, 400, 410, 445 and 500 s after ignition.

Figure 4-67. External Door View With Fan Prior to Ignition (0 s)

Figure 4-67 shows the external corridor doorway view prior to ignition. This was the only view that differs from the natural ventilation test due to the addition of the PPV fan. The difference can also be seen in the placement of the fan as explained in chapter one. The fan in the experiment was angled upward to cover the doorway. The fan in FDS was translated upward to cover the doorway as described in chapter 3.

Figure 4-68. Bunkbed View as Flames Involve Corner of Mattress (60 s)

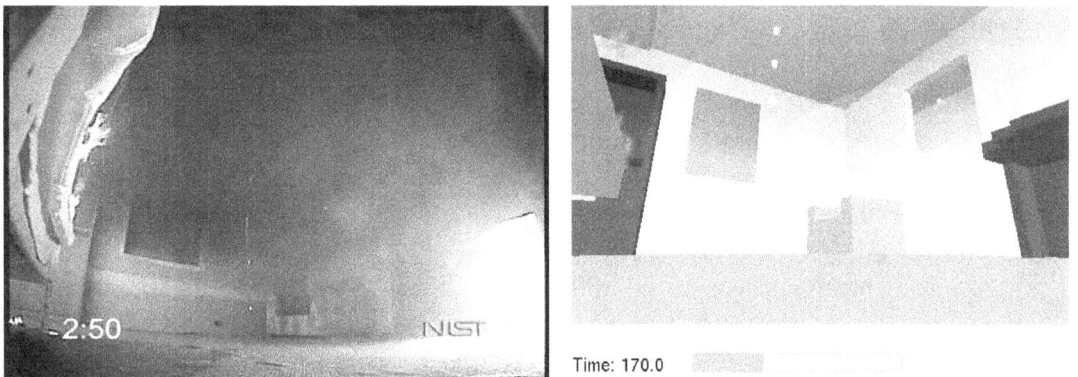

Figure 4-69. Doorway and Window View as Flames Spread to Top Mattress (170 s)

Figures 4-68 and 4-69 show two camera views during the growth stage of the fire. The bunk bed view shows the fire beginning and spreading on the corner of the mattress. The doorway and window view show the fire as it began to extend to the top bunk at 170 s. Some drop down of burning materials can be seen in the experimental frame of Figure 4-69; this was not captured in FDS and therefore increased the uncertainty of the simulation. The fire growth in this experiment was very comparable to the growth in the naturally ventilated experiment.

Figure 4-70. Flames Involving Bunk Bed (240 s)

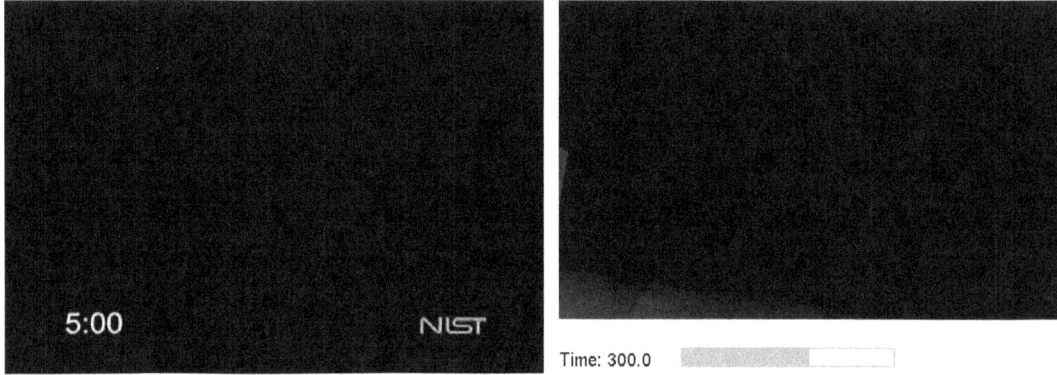

Figure 4-71. Visibility Lost in Bunk Bed View (300 s)

Figure 4-70 compares the fire development at 240 s after ignition. In both frames the fire is involving both mattresses that make up the bunk bed. The experimental growth was slightly ahead of that of the simulation. One reason for this could be the melted materials that have pooled below the bottom mattress in the experiment allowing for the bottom of the lower mattress to become involved in flames. The smoke was also denser in the experimental frame for the same reason.

Figure 4-71 has images captured 300 s after ignition. In both the experiment and the FDS simulation visibility had decreased greatly due to smoke filling and the bunk bed can no longer be seen. The naturally ventilated experiment lost visibility at approximately the same time.

Figure 4-72. Thick Smoke Flows From Corridor Doorway (300 s)

At the same time that visibility was lost in the room, Figure 4-72 shows the thick black smoke flowing from the corridor doorway. The smoke in the experiment appeared to be thicker than that of the simulation but both had smoke in the same regions of the doorways. Unlike the naturally ventilated simulation the positive pressure ventilated simulation did not over-predict the flame isosurface location.

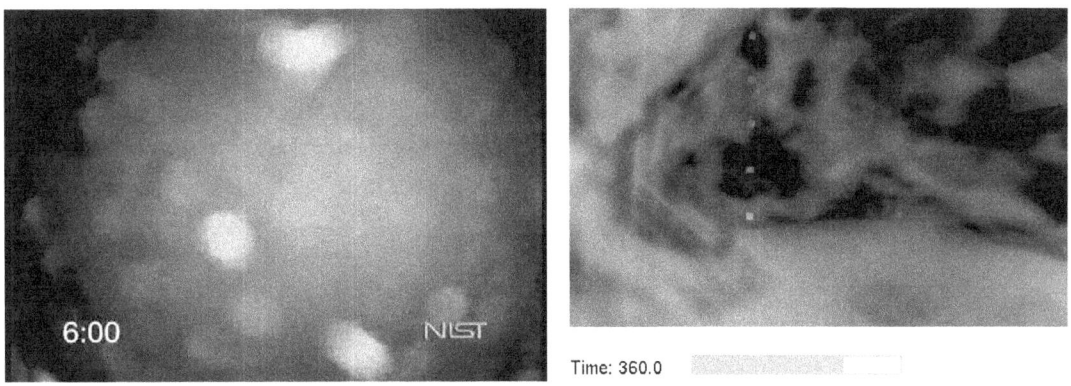

Figure 4-73. Doorway and Window View, From Inside the Room, Obstructed by Flames (360 s)

Figure 4-74. Doorway View 10 s After Fan is Turned On (360 s)

Figure 4-75. Flames From Window (360 s)

Figures 4-73, 4-74 and 4-75 were all captured at 360 s after ignition and 10 s after the fan was turned on. The internal view showed that the room was fully involved in flames. The experimental frame had some condensation on the lens which evaporated quickly. The external corridor doorway view displayed the ignition of the combustion products leaving the doorway both in the experiment and the simulation. The simulation under-predicted the amount of flames coming from the doorway. The final view at 360 s was an external view of the window. A large amount of flames were coming from the window in both scenarios. The flames filled the entire cross section of the window just as they did in the naturally ventilated scenario but due to the increased velocity added by the fan the flames extended outward away from the room more than in the naturally ventilated experiment.

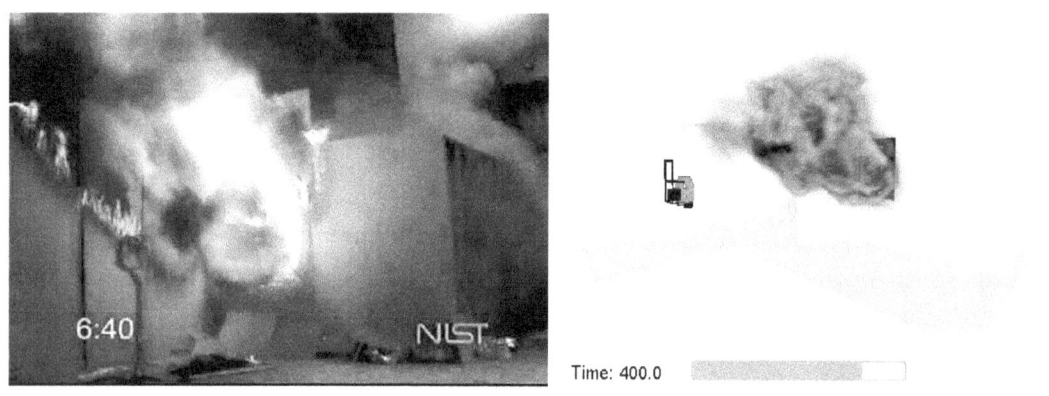

Figure 4-76. Flames Continue From Window (400 s)

Figure 4-76 has images taken of the window 55 s after the window was opened. Both images show flames filling the entire cross section of the window and exiting the window with force. Both images also have approximately the same amount of smoke being produced as the free burning from the window continues.

Figure 4-77. Combustion Products Forced into Room by Fan (410 s)

At 410 s the fan had nearly turned the flow of combustion products back into the room from the corridor doorway as shown in Figure 4-77. In both the experiment and simulation there was some smoke still coming from the top of the doorway but all burning was forced back into the room as can be seen from the glow in both images. This was evidence of the fans ability to reverse strong fire flows and FDS's ability to predict the phenomena.

Figure 4-78. Fire in Decay Stage (445 s)

Figure 4-79. Fan Forcing Flow Through Room (500 s)

Figures 4-78 and 4-79 show the fire in its decay stage. The experiment experienced its peak heat release rate although it continued to have burning outside of the room. The simulation frame showed significantly less combustion occurring outside the room. This was further evidence that it was difficult in FDS to create furniture that contained the same amount of fuel as actually exist. There was still burning in the room in the simulation but not of the magnitude of the experiment. This under-prediction of flames in the simulation was consistent with the under-prediction of temperatures in the room that was seen in the naturally ventilated scenario.

4.4.2.2 Numerical Comparisons

In this section, values of heat release rate, gas temperatures and gas velocities generated by the positive pressure ventilated FDS simulation will be compared to measurements from the full-scale experiments described in section 4.2.

Figure 4-80 compares the measured heat release rate leaving the experimental room to the heat release rate from the fire within the room as predicted by FDS. The experimental heat release rate peaked at 14 MW while the simulation heat release rate peaked at 16 MW. The difference in peak time corresponded to the time for the combustion products to leave the corridor doorway and window and reach the oxygen consumption calorimetry instrumentation in the exhaust hood. Both curves increased at a similar slope to the maximum output. This demonstrated that the sudden opening of the window was handled accurately by FDS. As with the naturally ventilated simulation the heat release rate curve in the simulation declined more rapidly than in the experiment. This was also due to the large uncertainty associated with the modeling of the rooms furnishings.

Figure 4-81 is a comparison of the naturally ventilated and positive pressure ventilated FDS simulations to isolate the fan's impact on the heat release rate. The naturally ventilated simulation peaked at 11 MW and the positive pressure ventilated simulation peaked at 16 MW, an increase of 45 %. This compared with the experimental increase of 60 % for the 200 s following ventilation. FDS also had convergence of the heat release rates of the two simulations as the fire burned down. This was based on the 100 s from 488 s to 588 s, when the PPV simulation terminated.

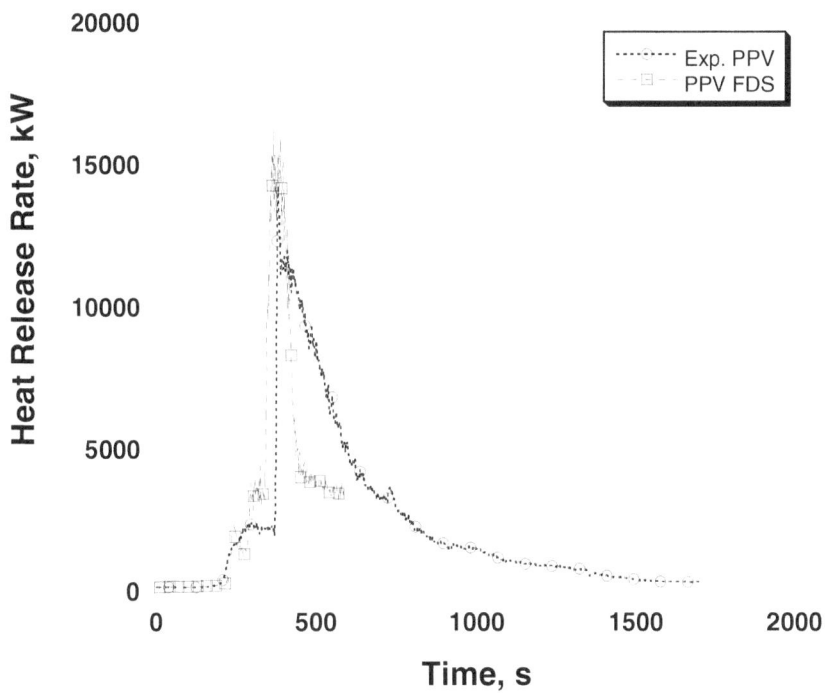

Figure 4-80. FDS and Experimental Positive Pressure Ventilated Heat Release Rate

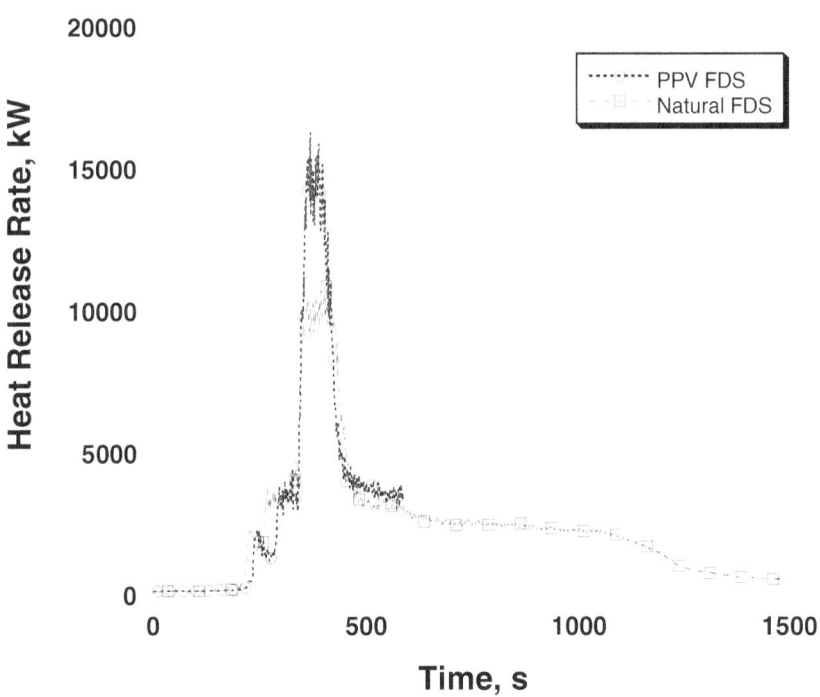

Figure 4-81. FDS Heat Release Rate Comparison of natural and PPV ventilation

The room gas temperatures in the FDS simulation, illustrated in Figure 4-82, increased similarly to those of the experiment in Figure 4-29 (on page 54). At 300 s both peaked at approximately 800 °C (1470 °F) at the ceiling prior to ventilation as the fire became oxygen limited. Once the window was opened the temperature in the simulation peaked 40 % (°C) higher than that of the experiment. At 500 s both the simulation and the experiment were 1000 °C (1830 °F) at the ceiling and decreasing at a similar rate. By the end of the simulation the ceiling temperatures decreased to 800 °C (1470 °F) in both cases. The lower level gas temperatures were lower than the experimental temperatures, partially due to radiation effects on the thermocouples as mentioned previously.

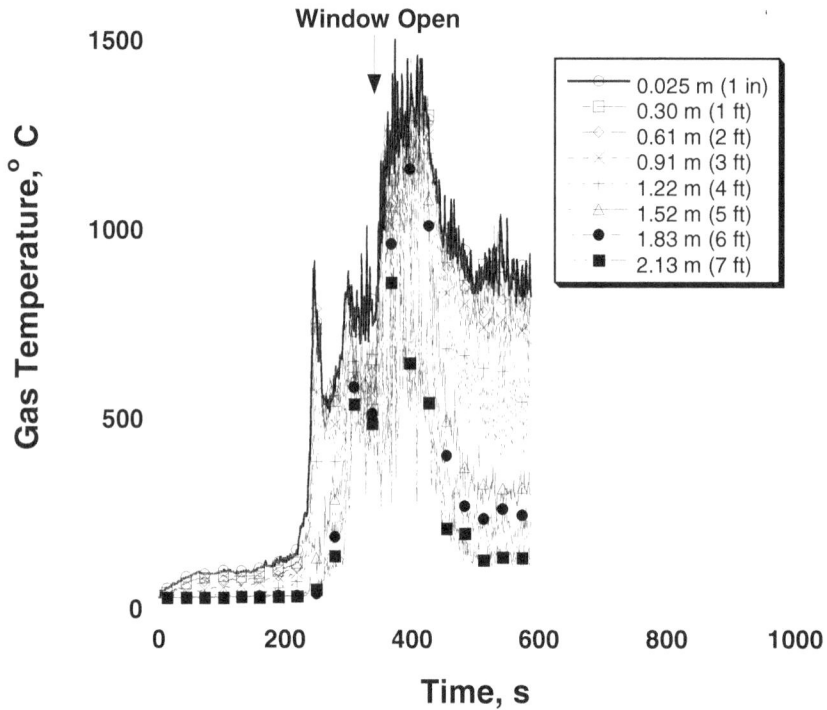

Figure 4-82. FDS PPV Ventilated Room Temperatures

Figure 4-83 presents the gas temperatures predicted in the fire room doorway. The simulation prediction points correspond to the measurements shown in Figure 4-31 (on page 69). The temperatures at all three points corresponded well between both cases. Top temperatures fluctuated between 600 °C (1110 °F) and 800 °C (1470 °F) prior to ventilation, peak at 1000 °C (1830 °F) to 1250 °C (2280 °F) after ventilation and decline to 600 °C (1110 °F) at the end of the simulation. Contrary to the previous lower level temperatures that were under-predicted, the temperatures in the lower half of the door in this simulation matched well with the experiment. This could be due to the decrease in radiation effects as increased amounts of ambient air was forced past the bare-bead thermocouples in the experiment.

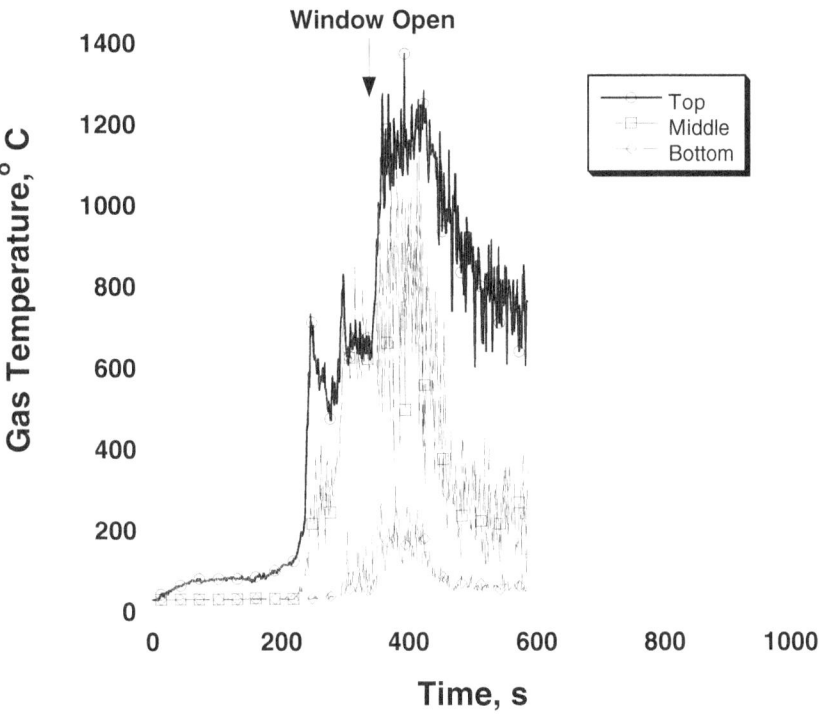

Figure 4-83. FDS PPV Ventilated Room Doorway Temperatures

Figure 4-84 shows the simulated corridor doorway temperatures that are compared to the experimental values in Figure 4-35 (on page 73). The temperatures in the top third of the doorway became greatly over-predicted as the window was opened and the temperatures peaked. For approximately 50 s after ventilation the simulated temperatures were double those in the experiment. This corresponds to the slightly longer time the fan in the simulation required to turn the flow around and out the window than occurred in the experiment. The lower level temperatures remained low in both cases as ambient air was being pulled directly in from the outside.

Figure 4-85 displays the gas velocities exiting the room out of the window. There was no bidirectional flow as the fan is turned on 5 s after the window was opened. These velocities are compared to the experimental velocities in Figure 4-39 (on page 64). Simulated velocities ranged from 6 m/s (20 ft/s) to 12 m/s (39 ft/s) at the time of ventilation as compared to experimental values of 10 m/s (33 ft/s) to 19 m/s (62 ft/s). In both cases the velocities decreased at a similar rate following the peak as the fire decreased in magnitude. The addition of the fan in the simulation caused a 3 m/s (10 ft/s) to 5 m/s (16 ft/s) increase in maximum velocity which was the same as the increase seen in the experiment.

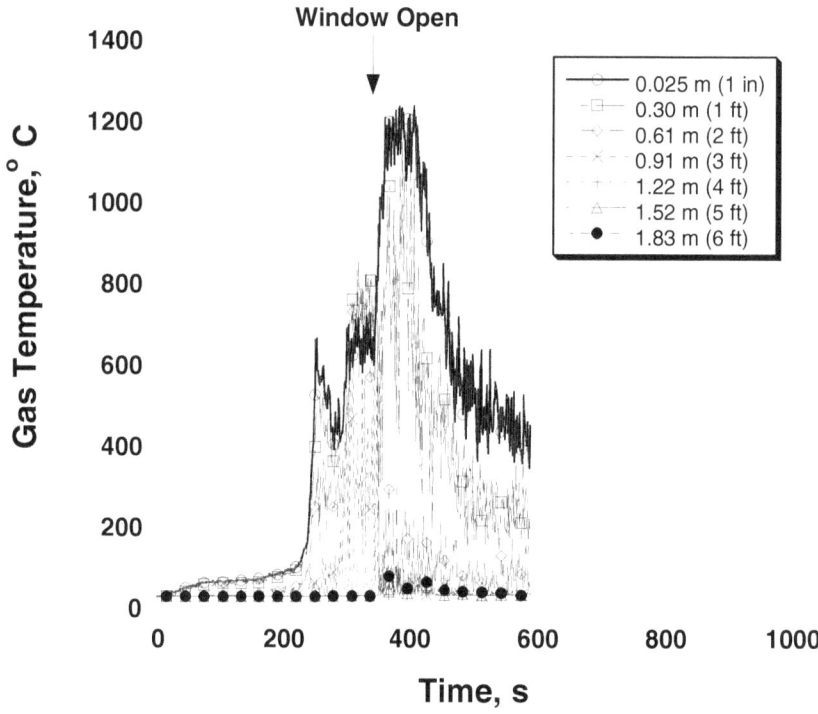

Figure 4-84. FDS PPV Ventilated Corridor Doorway Temperatures

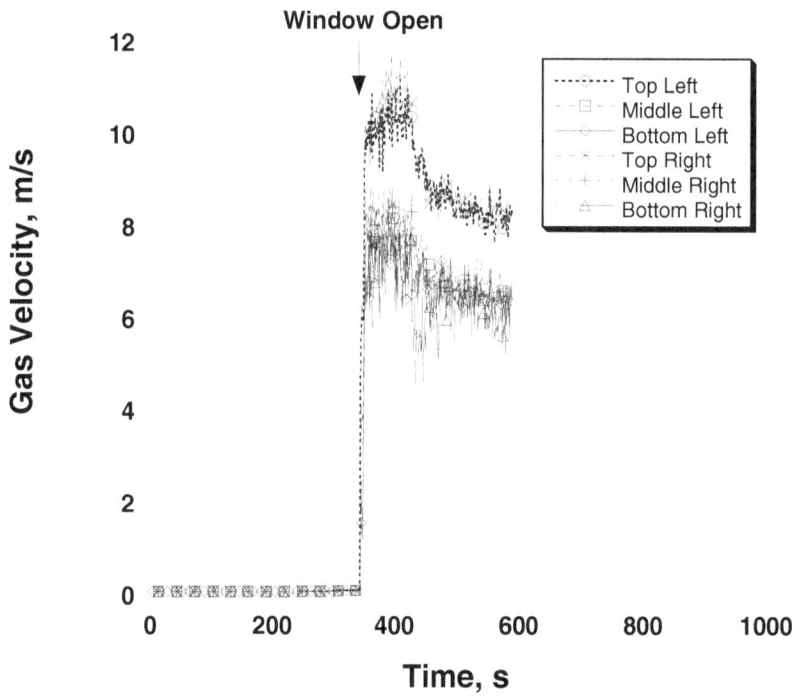

Figure 4-85. FDS PPV Ventilated Window Velocities

Figure 4-86 displays the gas velocities entering and exiting the room through the room's doorway. There was bidirectional flow as the fan was not able to completely reverse the fire flow at the plane of the doorway. These velocities are compared to the experimental velocities in Figure 4-41 (on page 66). Simulated velocities ranged

95

from -6 m/s (-20 ft/s) to 5 m/s (16 ft/s) at the time just after ventilation as compared to experimental values of -6 m/s (-20 ft/s) to 6 m/s (20 ft/s). In both cases the flow was out of the room at the top and into the room at the middle and bottom of the doorway. The addition of the fan in the simulation caused a 2 m/s (7 ft/s) to 4 m/s (13 ft/s) increase in maximum velocity which was the same as the increase seen in the experiment. This correlation was favorable for FDS's ability to capture the bulk effect of the PPV fan as the air is forced into the room.

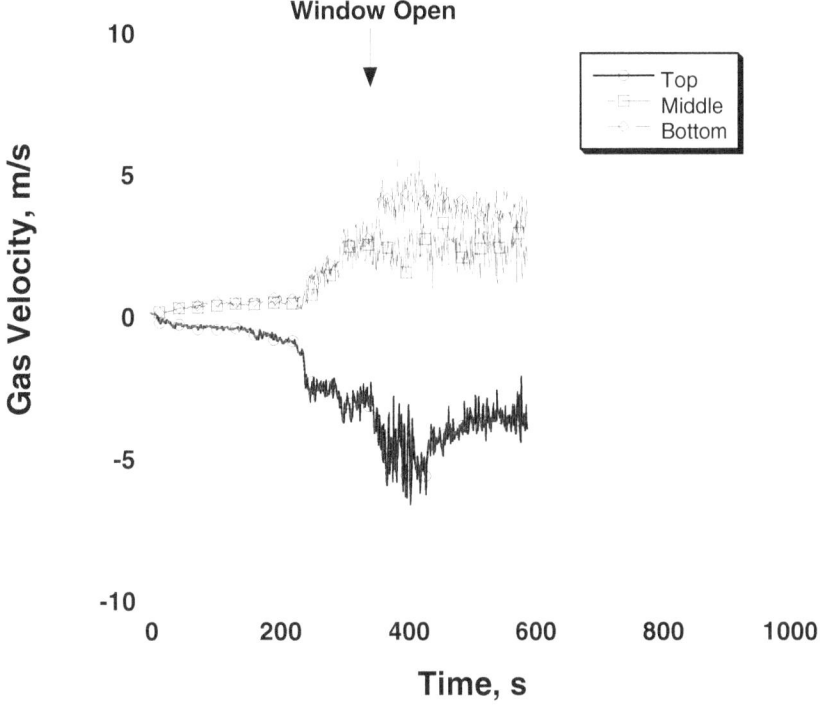

Figure 4-86. FDS PPV Ventilated Room Doorway Velocities

4.5 Discussion

Previous work has investigated the impact of positive pressure ventilation on gas temperature, window gas velocity, and mass burning rate by examining different fuel packages and different room/structural configurations. Stott studied typical home furnishings in a single story structure. Svensson utilized flammable liquid fires in three rooms of a larger structure. Ezekoye employed a polyurethane foam fuel in a 4 room residential structure. Different capacity PPV fans were utilized by each study. Instrumentation typically included upper and lower layer gas temperatures and one experimental series included pressure transducers for monitoring gas velocities.

Stott [15] conducted a series of experiments in Preston, U.K. utilizing furnished rooms. These experiments were instrumented with thermocouples but also relied upon subjective feedback from the participants that suppressed the fire or watched from outside. These experiments showed that there was a minor temperature increase after the use of the fan, approximately 10 °C (18 °F). The experiments also demonstrated that there was no flame extension to the corridor, the temperature decreased and the visibility improved after the initiation of PPV. The report also stated that the fire growth rate was not greatly increased by the fan. The increase in burning rate as PPV was initiated is consistent with this study for a single furnished room.

Svensson [16] of the Swedish Rescue Services Agency also conducted an experimental study of ventilation during fire fighting operations. His experimental setup utilized three rooms on the first floor of a fire training facility. A 0.5 m (1.6 ft) diameter heptane pool fire was utilized which generated a heat release rate of 0.37 MW. This was a significantly smaller fire (in terms of heat release rate) than the fully furnished rooms in these experiments which released heat at a rate of 11 MW to 14 MW. The smaller heat release rate heptane pool fire generated much lower temperatures, 300 °C (570 °F), than was monitored in the current study, 800 °C (1470 °F). Svensson did report an increase of 40 % in the burning rate after PPV was initiated. This is comparable to the 60 % increase that was produced in the furnished room in this study. The furnished room had a greater fuel surface area which would have been consistent with the difference in the burning rate. Svensson also reported significantly lower pressure differentials. The larger room size may account for some of this difference but the main reason is likely the smaller fan output. The fan used in Svensson's experiments was rated at approximately one third that of the one used in these experiments. This difference in fan size is also consistent with the smaller flows that were recorded in Svensson's experiments.

Another set of experiments was conducted in the United States by Ezekoye, et al[17] of the University of Texas at Austin. These experiments examined positive pressure attack for heat transport in a house fire. The fuel source chosen for those experiments was 9 kg (19.8 lb) of polyurethane foam, oriented on a rack, capable of generating a peak heat release rate per unit area of 1.2 MW/m^2. This fuel package produced temperatures of 760 °C (1400 °F) at the ceiling and 200 °C (390 °F) at the lower levels but did not appear to cause post flashover conditions that were present in the furnished room fire experiments. The fan flow rate was similar to the one used in the furnished room. The mixing which caused higher temperatures in the lower layer are seen in both sets of experiments. Those experiments only reported temperatures so pressure differential, burn rate and gas flow velocity could not be compared.

There has been very little documented computational fluid dynamics modeling of positive pressure ventilation. The only other documented FDS modeling is briefly explained in the University of Texas report [17]. In these simulations the fan was prescribed as a single vent at the face of the door. These simulations did not examine the effects of the fan on the fire and only looked at the temperatures in a victim room. The authors reported that the FDS simulations created similar trends but there was no figures displaying the correlations.

Another CFD model, SOFIE, was used by Gojkovic and Bengtsson [18] in Sweden to examine the possibility of backdraft conditions. A three room setup was used to examine the possible effects of PPV both correctly and incorrectly on an under-ventilated fire. The fan was prescribed as an inflow boundary in the front door. The authors stated, "an inflow boundary does not simulate the characteristics of the fan very well." The authors also concluded that the PPV causes increased mixing and increased chance of backdraft but for a short duration.

4.6 Room Fire Summary

Compared to natural ventilation, positive pressure ventilation caused lower fire room temperatures, increased window gas flows and higher pressure differentials for this set of furnished room burns. After the peak heat release rate was reached in both experiments, the temperatures with PPV remained 200 °C (360 °F) to 400 °C (720 °F) below the temperatures with natural ventilation for 800 s. The doorway temperatures with PPV peaked 200 s before, but quickly dropped to 200 °C (390 °F) to 500 °C (930 °F) below the naturally ventilated temperatures for 900 s following the peak in the PPV experiment. The window gas temperatures generated during the PPV experiment peaked 200 seconds before gas temperatures at the same location in the naturally ventilated experiment. In the naturally ventilated experiment, gas temperatures in the top two thirds of the window were higher than the PPV experiment, but gas temperatures in the bottom third were lower due to the inflow of air. Once the fan was running, the PPV corridor temperatures were as much as 500 °C (930 °F) less than comparable temperatures in the naturally ventilated experiment.

The PPV fan alone generated gas velocities of 5 m/s (16 ft/s) in the window while the naturally ventilated fire generated velocities of nearly 12 m/s (39 ft/s). In the experiment with the PPV fan, window gas velocities of nearly 20 m/s (66 ft/s) were generated, approximately equal to the additive velocities from the fan and the naturally ventilated fire. The fan quickly forced a unidirectional flow out of the window but took a period of time to completely reverse the flow out of the doorway and create a flow into the room. The fan was able to create a more tenable atmosphere as soon as it was turned on by reversing the natural flow out of the corridor, where the fire fighters would be approaching the fire for extinguishment.

The heat release rate of the fire was increased by the fan for the 200 s following the peak heat release rate. This was critical because this is the time period during which the fire department would typically be advancing to extinguish the fire. The peak heat release rate for the two experiments occurred at approximately the same time and the rate with the PPV fan was 2 MW higher. The PPV fan caused a 60 % increase in burning rate during this time of initial fire department attack. This reinforces the importance of selecting a ventilation location close to the seat of the fire that allows for all of the combustion products to be ventilated to the exterior of the structure. The PPV ventilated experiment forced the flames at least 1.83 m (6 ft) out of the room as compared to the 0.91 m (3 ft) by the naturally ventilated experiment. Flame extension out of the building openings may also pose a potential ignition hazard to materials nearby.

While the use of PPV in this particular configuration caused an increase in the room's fire burning rate, it lowered the temperatures in the room, forced all of the combustion products to flow out of the room without affecting the corridor and improved the visibility leading up to and in the room itself. In this experimental configuration, a fire fighting team would likely have been able to attack the PPV ventilated fire more easily than the naturally ventilated fire.

This limited data set indicates that coordination of fire fighting crews is essential to carry out positive pressure ventilation in the attack stages of a fire. In this experiment, ideal coordination was simulated as the window was ventilated in the correct location and the fan was initiated seconds later. Once the fan was turned on, it took approximately 60 s to 90 s for the fire to reach its peak burning rate and for the flow to be forced away from the entrance. After this transition, the fire remained at a steady burning rate until the fuel was consumed. This would indicate that for the conditions in this experiment fire fighters should delay 60 s to 120 s after ventilation and fan start before advancing towards the fire. This would allow the flows to stabilize, temperatures to decrease and visibility to improve. The burning rate of the fire could become steady at the rate determined by the modified air flow and would be less likely to change rapidly as the fire fighters approach. The time to reach this new steady condition could vary with building layout, fire size, fuel load and fan capacity.

The visual and numerical comparisons demonstrate that the fire behavior of a room fire both with and without positive pressure ventilation can be modeled by FDS and visualized with Smokeview. Some differences exist in the geometry and material properties, but the fire dynamics and the net impact of the positive pressure ventilation fan can be captured with an acceptable degree of accuracy.

The heat release rate, gas temperature and gas velocity comparisons show reasonable agreement between the experiments and the model in terms of both trends and range. Significant differences were displayed in the gas temperatures, partially due to radiative effects. Future simulations could utilize the FDS prescription "THERMOCOUPLE" identification of measurement points instead of "temperature" to better compare the simulation results to experimental results by incorporating the radiative effects in the model as well. The two experiments simulated using FDS, reproduced the effects of a positive pressure ventilation fan on a post flashover room fire, and support the use of the model fan on larger scale scenarios that may not be able to be supported by full scale experimental data.

Chapter 5: Colonial House Practical Scenario

5.1 Scenario Overview

In order to expand the understanding of the effects of positive pressure ventilation, it would be very informative to perform experiments in various types of full-scale structures. These structures are very difficult to obtain and very expensive to instrument. Using the lab scale tests documented above as a calibration for the FDS, it is possible to visualize the effects of positive pressure ventilation applications in full-scale scenarios.

A colonial house was chosen due to the potential for many fire fighting scenarios. The house had two floors and a basement. The first floor was made up of a study, living room, dining room, kitchen, bathroom, laundry room, sun room, family room and garage. The second floor had four bedrooms and two bathrooms. The basement had an open floor plan and was unfinished.

Two scenarios were examined with a fire in the rear bedroom of the second floor. The fire started next to the bed to simulate a fire that originated in a trashcan. The doors to all of the rooms on the floor were open and the fire grew without intervention for the first 240 s. At 240 s in both scenarios the front door was removed (opened) to simulate the fire department's arrival at the front door. Five seconds after the front door opened, the window to the fire room was removed to simulate the fire department venting the fire room. One simulation allowed the fire to behave without fire department intervention, and in the second scenario a PPV fan located at the front door was activated at 250 s. The purpose of the two scenarios was to analyze the impact of the addition of the PPV fan on the fire conditions.

The simulations were run for 800 s and provided insight into the potential fire development and spread as well as the potential impact of the positive pressure ventilation implementation. Issues that were addressed include the fans' effect on fire growth, smoke spread, temperatures, oxygen concentrations and velocities in the pathway to the fire room potentially occupied by fire fighters and tenability criteria of adjacent rooms to the fire room that could be occupied by victims of the fire.

5.2 Computer Simulations

5.2.1 Domain

The computational domain used for this analysis measured 16.4 m (53.8 ft) wide x 13.9 m (45.6 ft) deep x 10.0 m (32.8 ft) tall. The computational domain was divided into 3 grids (Figure 5-1). The grid containing the basement (grid 1) measured 16.4 m (53.8 ft) wide x 13.9 m (45.6 ft) deep x 2.5 m (8.2 ft) tall. The grid containing the first floor (grid 2) measured 16.4 m (53.8 ft) wide x 13.9 m (45.6 ft) deep x 3.125 m

(10.3 ft) tall. The grid containing the second floor (grid 3) measured 16.4 m (53.8 ft) wide x 13.9 m (45.6 ft) deep x 4.4 m (14.4 ft) tall. All three grids were composed of 0.15 m (6 in) cubic grid cells. The domain contained a total of 636,000 grid cells (Figure 5-2).

The use of 3 grids or "multiple meshes" allowed the simulations to be parallel processed. The three meshes were processed on three separate CPU's which greatly reduced computational time. Each mesh contained the same size grid cells which allowed for optimum sharing of data from mesh to mesh. The governing equations were solved with a time step based on the fire plume spread rate in the particular meshes. Due to the fact that each mesh could have different time steps, this saved CPU time by updating the meshes only when necessary. When the fan was added for the PPV simulation, the boundaries to its domain were also open and the vents for the fan were located as described in chapter 1 (Figure 5-3).

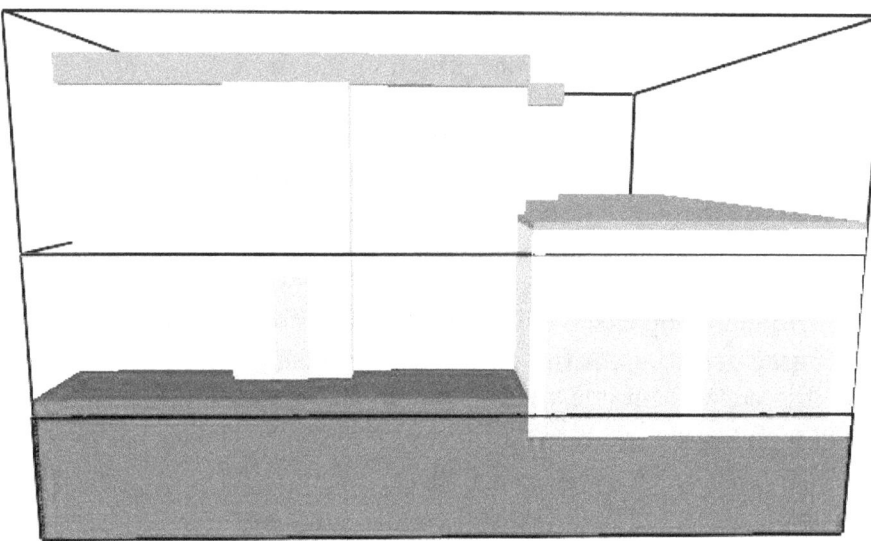

Figure 5-1. Colonial House and Grid Locations

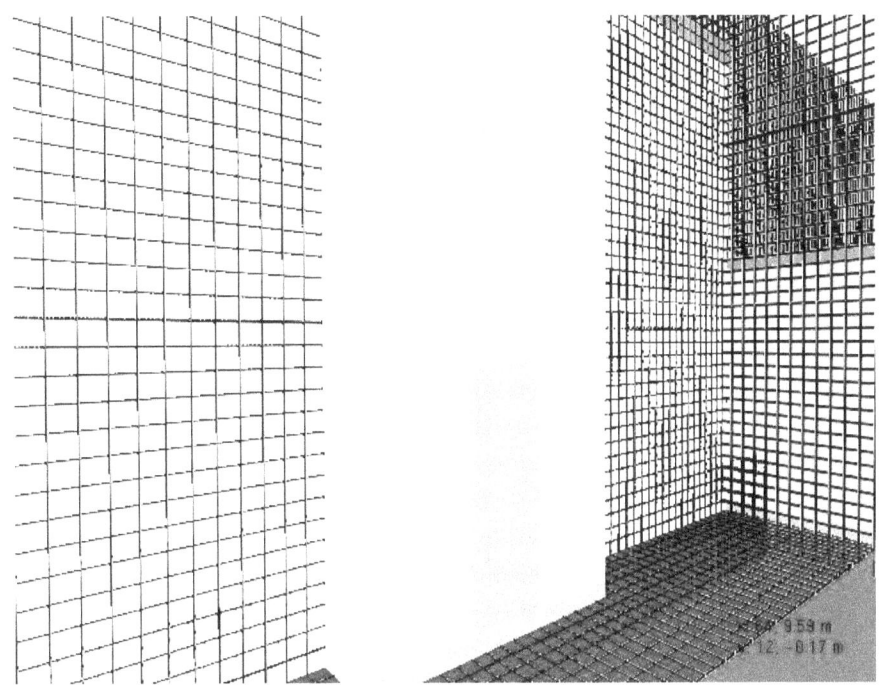

Figure 5-2. Display of Grid Cell Size

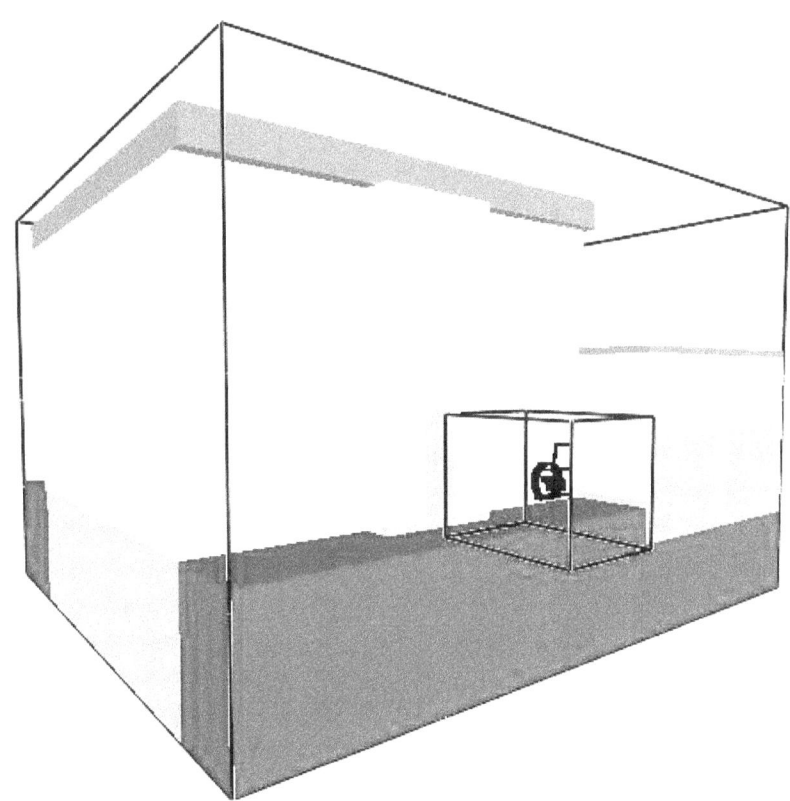

Figure 5-3. Colonial House and PPV Fan Placement

5.2.2 Geometry

The floor plans of the house are shown in Figures 5-4 through 5-11. The size and location of the walls, doorways, windows and furniture were based on the floor plans. All of these obstructions were adjusted by FDS to correspond to the nearest computational cell location. This resulted in objects, used in the model, that appear thicker then they would be in reality.

Figure 5-4. Floor Plan of First Floor

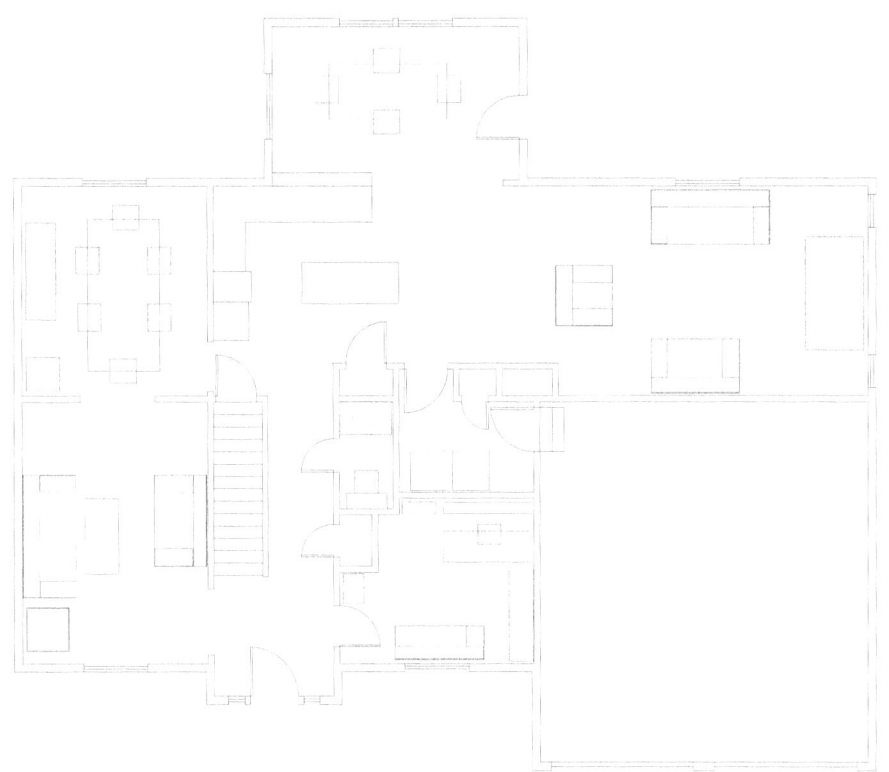

Figure 5-5. Furniture Locations on First Floor

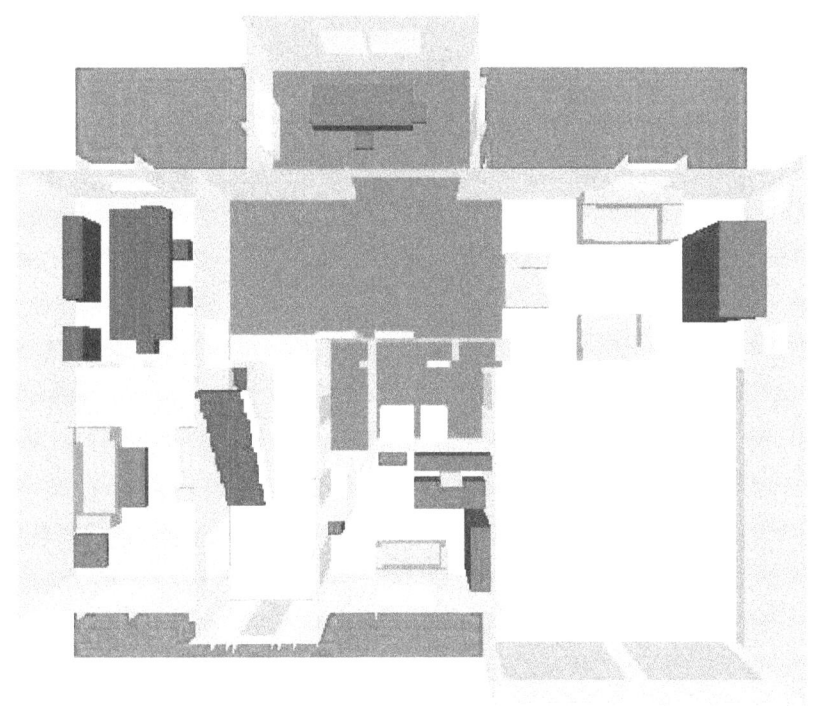

Figure 5-6. Smokeview Display of First Floor

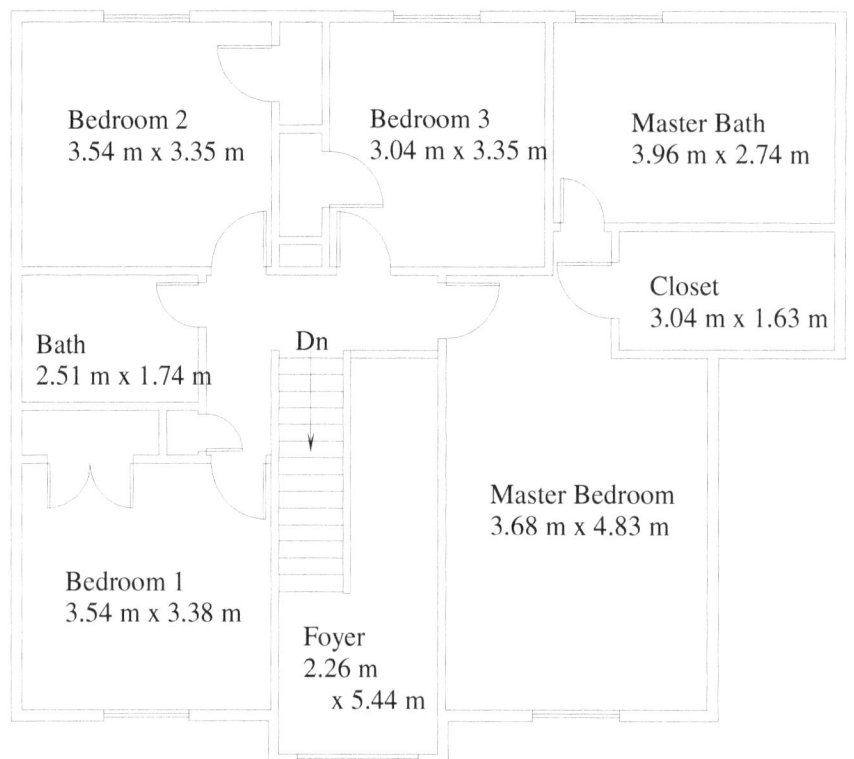

Figure 5-7. Floor Plan of Second Floor

Figure 5-8. Furniture Locations on Second Floor

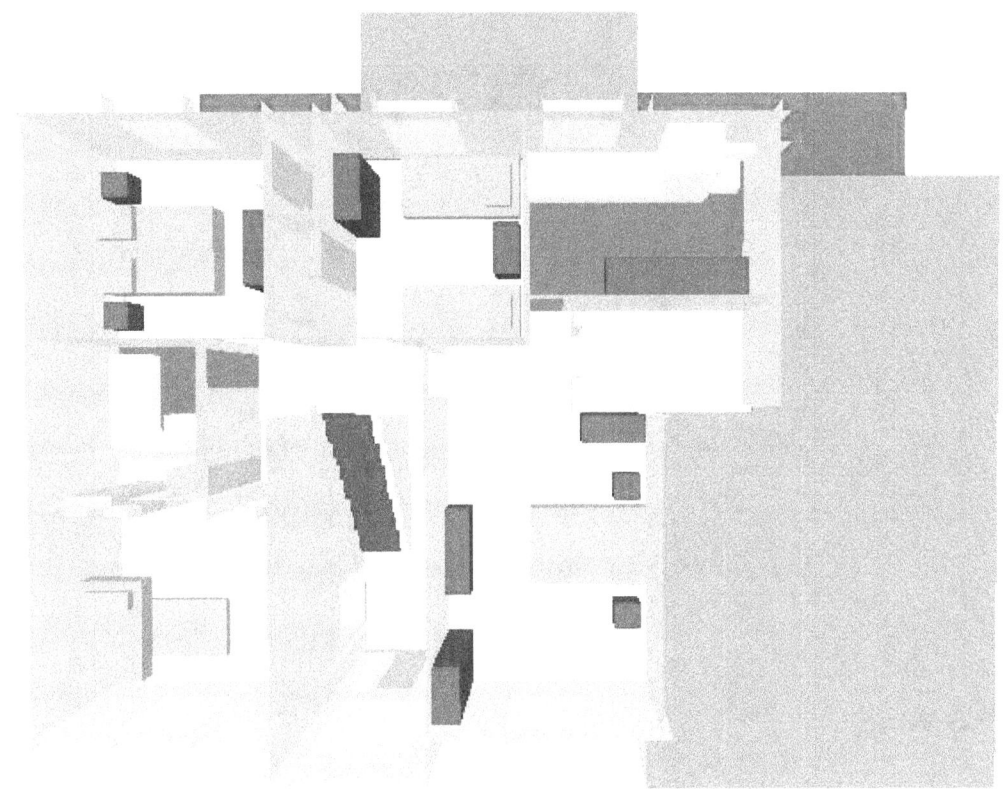

Figure 5-9. Smokeview Display of Second Floor

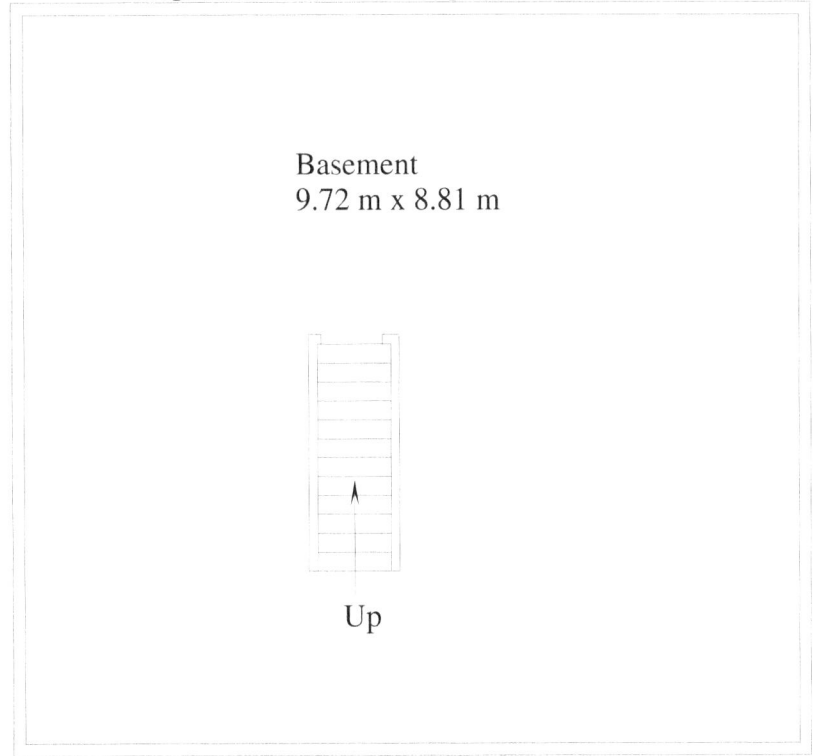

Figure 5-10. Basement Floor Plan

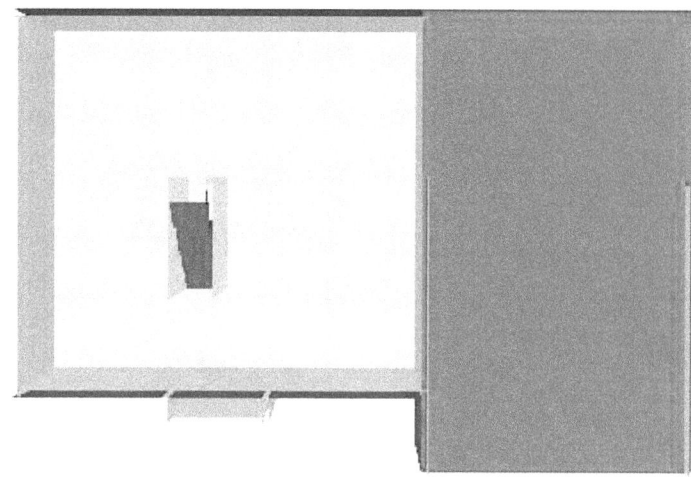

Figure 5-11. Smokeview Display of Basement

5.2.3 Vents

This simulation considered seven vents or openings from the structure to the outside, the front door, the window of the fire room, and five small openings to the roof area. The front door was 0.76 m (2.5 ft) wide x 2.0 m (6.7 ft) high. This vent was opened at 240 s in both scenarios. The window of the fire room measured 1.1 m (3.5 ft) wide x 1.25 m (4.1 ft) high with a 0.8 m (2.6 ft) sill height (Figure 5-12). This vent was opened at 245 s in both scenarios. The small openings in the second floor were placed in all four bedrooms and in the master bathroom to simulate air vents and leaks to the roof area. The front and rear edges of the roof area were open to the exterior of the house to simulate the eaves and leaks from the roof. The small openings to the roof measured 0.3 m (1.0 ft) long x 0.15 m (0.5 ft) wide (Figure 5-13). The openings along the edges of the roof were 0.3 m (1.0 ft) wide. The openings in the ceiling and the roof edges were also open over the entire duration of the simulation.

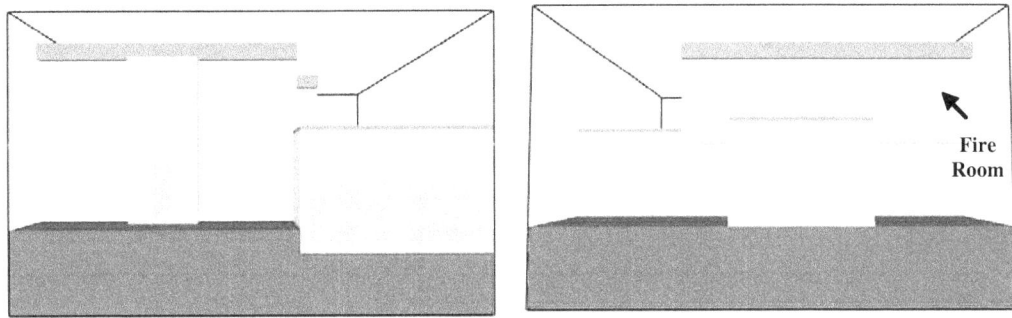

Figure 5-12. Front and Rear View of House

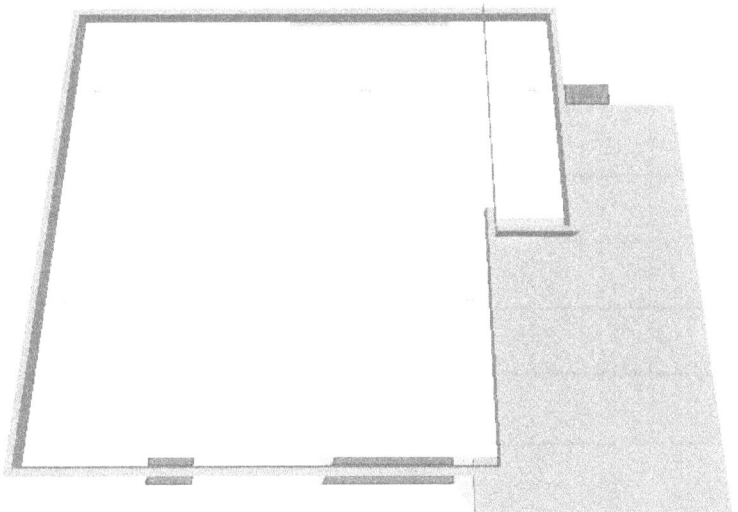

Figure 5-13. Location of Room Vents with Roof Removed

5.2.4 Materials

Each obstruction in the domain was given a set of physical and thermal properties that were used in the calculation. Each wall, ceiling, floor and piece of furniture is defined with properties such as ignition temperature, heat of vaporization, density and thickness (see Table 9). All of the walls and the ceiling of the house were gypsum board and the floor was covered in carpet. The primary fuels in the house were upholstery and oak. Figure 5-14 displays a view from inside the front door. Many materials can be identified, the brown steps were oak, the white walls were gypsum board, the light blue floor was carpet, the orange floor was also carpet (aesthetic purposes only) and the light blue couch was upholstery. All of the material properties were derived from the FDS database [6].

Table 9. Colonial FDS Input Material Properties

Material	Ignition Temperature (°C)	Thickness (m)	Heat Release Rate Per Unit Area (kW/m^2)	Heat of Vaporization (kJ/kg)
Upholstery	280		1700	
Oak	340	0.02		4000
Plastic	370		500	
Carpet	280		3000	
Gypsum Board	400	0.013	100	
Sheet Metal	N/A	0.0013	N/A	N/A
Pine	390	0.02		2500
Concrete	N/A	0.2	N/A	N/A
Glass	N/A	0.005	N/A	N/A
Thin Oak	340	0.005		4000
Floor Tile	280			3000
Grass	N/A	1.0	N/A	N/A

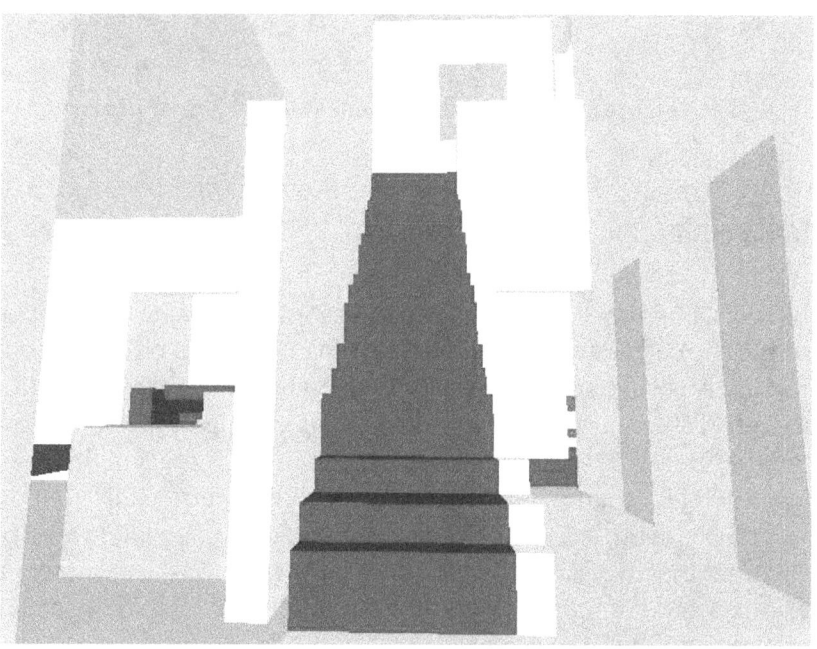

Figure 5-14. View of Interior from Inside the Front Door

5.2.5 Fire Source

A vent measuring 0.15 m x 0.15 m (0.5 ft x 0.5 ft) was used for ignition and was located between the bed and the night stand in the rear bedroom on the second floor. The vent had a defined heat release rate per unit area of 2150 kW/m^2 to simulate a 50 kW trashcan fire. The vent was placed on top of a 0.15 m (0.5 ft) cube to simulate a small trashcan. This vent produced fuel for the duration of the simulation (Figure 5-15).

Figure 5-15. Fire Source in Bedroom

5.2.6 Output Files

Output files were prescribed to best analyze the fan's effect on fire intensity in and around the fire room. In addition, smoke movement, temperatures, oxygen concentrations, and velocities in the pathway from the front door to the fire room which potentially would be occupied by fire fighters were analyzed. Tenability criteria of adjacent rooms to the fire room that could be occupied by victims of the fire were evaluated as well. Vertical and horizontal temperature, oxygen and velocity slices were placed through the center of each room and many of the openings. Heat release rate was continually recorded in the domain as well as the flame isosurface locations. Three dimensional smoke was also used for analysis of smoke movement and visibility. Comparisons of these output files between the two scenarios are located in Section 5.3.

5.3 Results

5.3.1 Fire Growth and Smoke Spread

Images were rendered from the naturally ventilated and positive pressure ventilated FDS simulations to characterize the effects of the positive pressure ventilation tactic on fire growth and smoke spread. Iso-surfaces of the heat release rate per unit volume and three-dimensional smoke density parameters are displayed in Figures 5-16 through 5-20.

Figure 5-16 shows the fire growth and smoke spread prior to ventilation. These images from 60 s to 240 s were identical for both ventilation scenarios up to 240 s. At 60 s the fire was growing between the night stand and the bed. Smoke was beginning to layer at approximately 1.2 m (4 ft) from the floor. The flames began to attach to the bed at 120 s and the smoke layer thickened. At 180 s the flames were spreading across the bed and the flames appeared to reach the ceiling. The smoke continued to thicken and the layer descended to 0.9 m (3 ft) from the floor. By 240 s the smoke layer had reached the floor and visibility was significantly decreased in the bedroom of origin. The view was moved to just inside the front door at 240 s. This frame shows the smoke almost reaching the floor in the hallway at the top of the stairs to the second floor. Another view from the exterior at 240 s displays smoke coming from the eaves of the roof on the front side of the house.

Figures 5-17 through 5-20 show comparisons of the two ventilation techniques; natural ventilation on the left and positive pressure ventilation on the right. Figure 5-17 is the comparison five s after ventilation of the rear of the house. Both frames have smoke coming from the window and eaves of the roof. At this point there was not much of a difference between the two. Figure 5-18 shows the front of the house at 360 s. The naturally ventilated frame had smoke flowing from the eaves across the whole front of the house. The PPV ventilated frame had increased flow out of the eaves due to the increased flow from the fan. Figure 5-19 is a view of the rear of the house at the same time as the previous Figure. At this time there were flames coming out of the entire cross section of the ventilation window. Both frames had about the same amount of flames with the PPV frame having the flames further out of the frame as opposed to the flames in the naturally ventilated flame that go straight upward. This is another sign of the increased flow created by the PPV fan. The final comparison in Figure 5-20 shows the fire in the decay stage at 600 s. Flames were no longer projecting from the windows in either frame but the amount of smoke being forced out the window was greater in the PPV simulation.

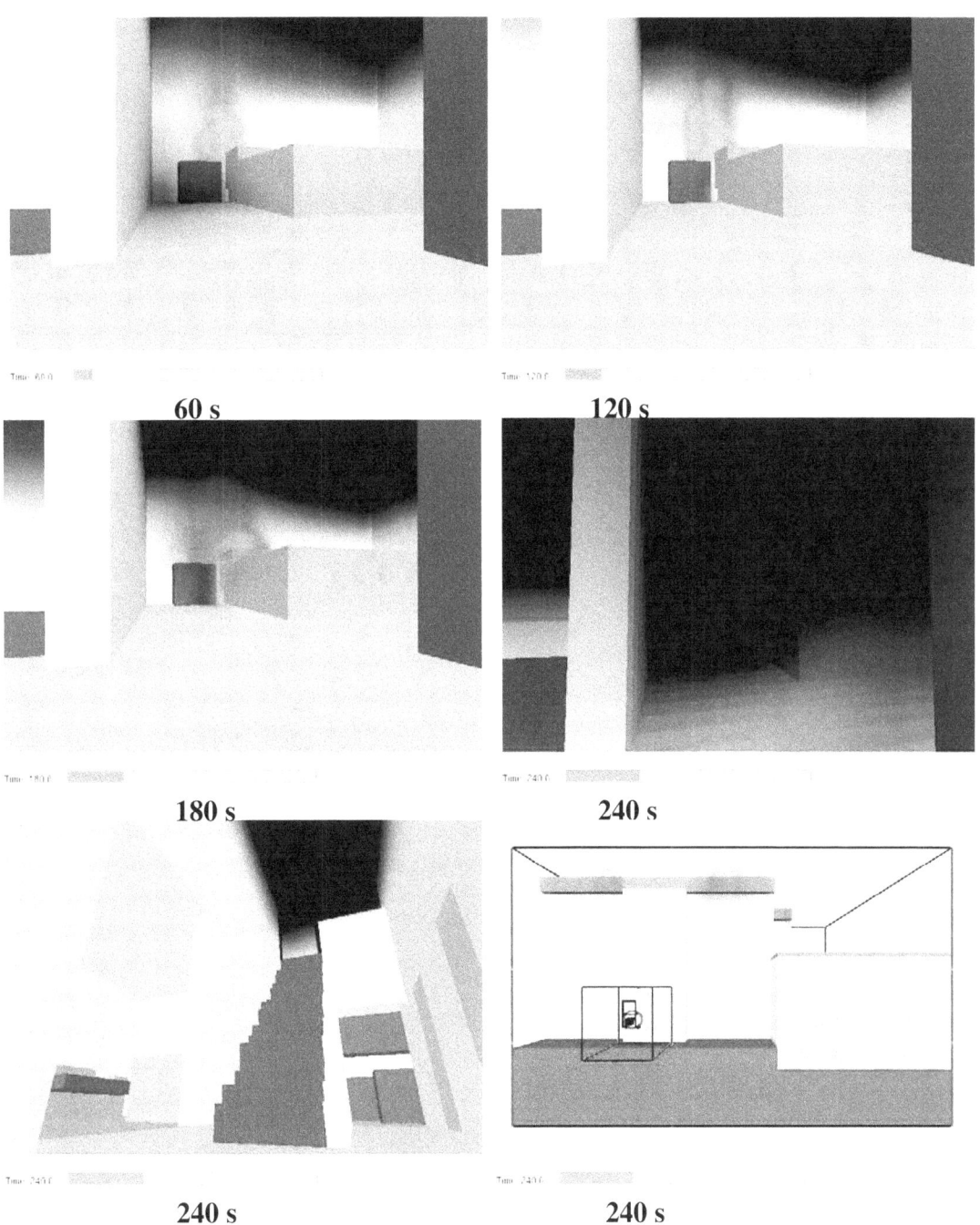

Figure 5-16. Growth of Fire Prior to Ventilation, 60 s to 240 s

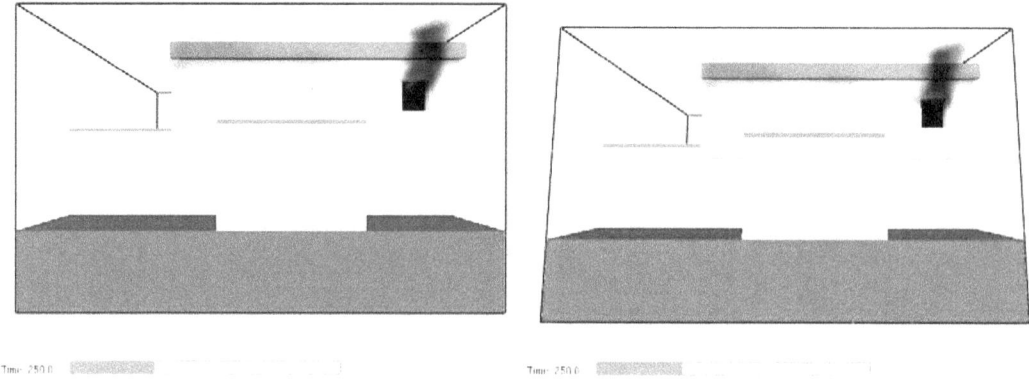

Figure 5-17. Comparison of Simulations at 250 s (Natural left, PPV right)

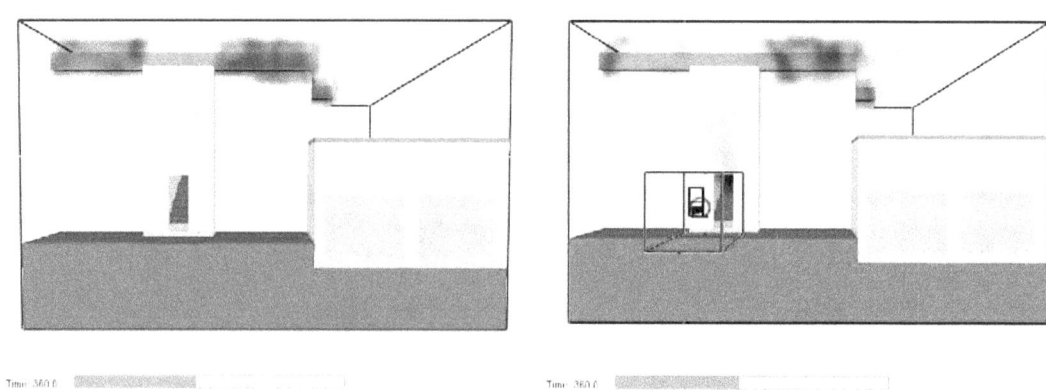

Figure 5-18. Comparison of Simulations at 360 s (Natural left, PPV right)

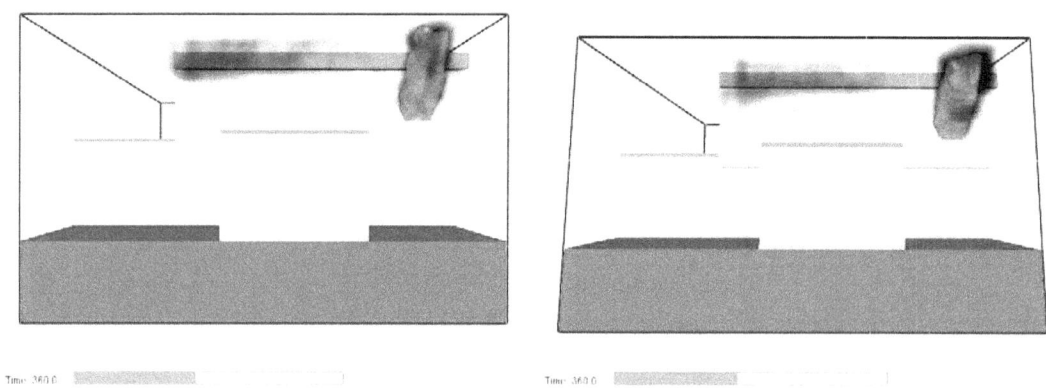

Figure 5-19. Comparison of Simulations at 360 s (Natural left, PPV right)

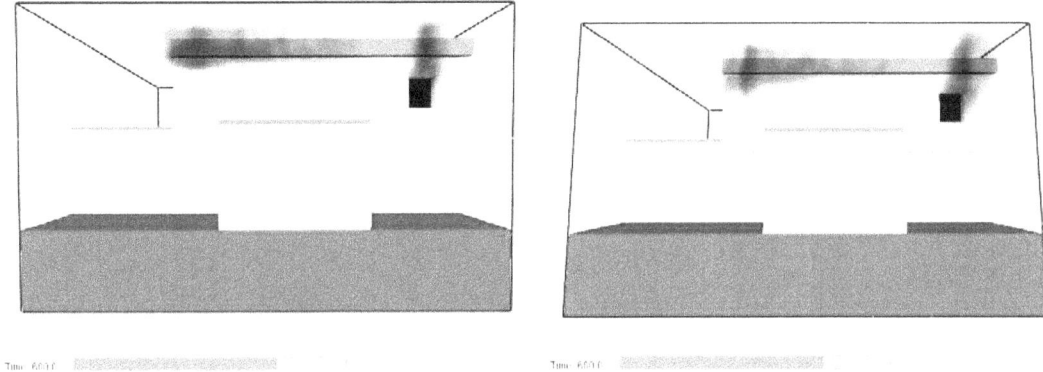

Figure 5-20. Comparison of Simulations at 600 s

5.3.2 Heat Release Rate

The heat release rate for the duration of both simulations is plotted in Figure 5-21. As expected the two curves are identical for the first 250 s prior to ventilation. After ventilation the naturally ventilated simulation peaks approximately 25 s prior to that of the PPV simulation. The reason for this may be due to the proximity of the vent window to the seat of the fire. The naturally ventilated fire was able to pull fresh air from the window while the forced flow of the PPV fan caused the flow to be out of the window and the source of fresh air came from the flow in through the front door of the house. This was seen in the room fire experiment but at a lower magnitude due to the lack of a long distance between the fire room and the fan.

The heat release rate in both simulations peaked at approximately the same value of 7.3 MW. This is not consistent with the room fire scenario but could also be due to the increased distance between the fire room and the PPV fan. The curves were consistent with the experiments after the peak was achieved. The positive pressure ventilated scenario maintained an elevated heat release rate of approximately 3 MW higher for the 70 s following the peak output. This is consistent with the room fire comparison which also demonstrated a comparable heat release rate differential for a similar period of time. After this 70 s of increased burning, the two curves converged, similar to the experimental results. The sudden drop in the natural ventilation scenario was due to a transition to a ventilation limited state in the fire room. The forced ventilation of the fan caused more air to reach the fire and minimize the sudden drop in heat release rate.

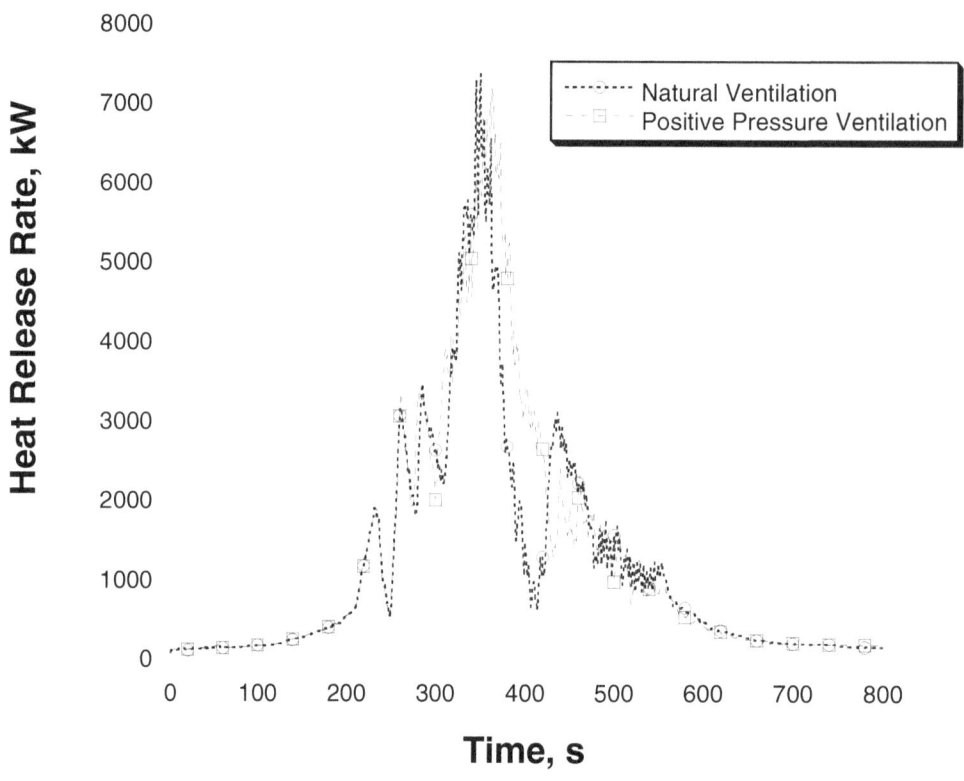

Figure 5-21. Colonial House Heat Release Rate Comparison

5.3.3 Temperature

Temperature slices were examined to assess the tenability conditions that existed for the duration of the fire. Horizontal slices were taken at 1.2 m (4 ft) above the floor with the geometry above the slice rendered transparent to examine the temperature distribution throughout the floor of the house. Threshold temperatures were chosen for victims and firefighters.

Research by Montgomery [19] in 1975 indicated that in humid air rapid skin burns would occur at 100 °C (212 °F), and 150 °C (300 °F) was the exposure temperature at which escape was not likely. In 1947, Moritz [19] experimented on large animals and found that 100 °C (212 °F) represented the threshold for local burning and hyperemia (general burning). For this analysis, a temperature value of 100 °C (212 °F) was considered the temperature at which victims could be incapacitated.

Fire fighters operating in structures were also susceptible to injury from temperature exposures. Fire fighter protective clothing standards such as NFPA 1971 refer to exposures of 260 °C (500 °F) for five minutes [20]. This same temperature was the minimum temperature to which the thread that holds the garments together must endure. Other data exists that a firefighter can survive flashover conditions of 816 °C

(1500 °F) for up to 15 s [27]. For this analysis, a temperature value of 300 °C (572 °F) was considered the temperature at which firefighter's turnout gear will begin to degrade and it is no longer safe for the fire fighter to remain in the atmosphere.

Figures 5-22 and 5-23 show the second floor temperatures 1.2 m (4 ft) above the floor in 100 s intervals for 600 s. At 100 s the fire is still in the growth stage and the temperatures in both scenarios remained below 100 °C (212 °F). The temperatures increased to approximately 130 °C (266 °F) at 200 s in both scenarios which was consistent with the fact that ventilation had not taken place. At 200 s it can be estimated that victims in the fire room would be incapacitated but fire fighters could still operate safely. All other rooms on the second floor remained thermally tenable by both victims and fire fighters.

By 300 s the window and front door had been opened for 60 s and the fan had been flowing for 50 s. In both scenarios the fire room temperatures had reached those consistent with flashover and firefighters could no longer operate in the fire room. The PPV ventilated fire room had lower temperatures than the naturally ventilated scenario. At this time, a difference in temperature was obvious in the rooms left of the stairwell and the stairwell itself. The flow created by the PPV fan had kept the temperatures near ambient while the naturally ventilated temperatures were reaching or had reached the tenability limit for victims.

The fire had reached its maximum output by 400 s and had heated a good portion of the second floor. The fire room temperature of the PPV ventilated scenario was higher than that of the naturally ventilated scenario but the rest of the second floor was cooler in the PPV scenario. In the naturally ventilated scenario the tenability criteria for victims was reached in all of the second floor rooms, but the master bathroom in the remote corner from the fire. Temperatures reached as high as 200 °C (392 °F) in the stairwell which was not above the fire fighter temperature threshold but it was in the path they would take to attack the fire, thus making advancement more difficult. In the PPV ventilated scenario, the rooms to the left of the stairwell exceeded the victim incapacitation temperature but the right half of the floor and the stairwell remained thermally tenable for both victims and fire fighters.

The fire continued through its decay stages in the 500 s and 600 s comparisons. The naturally ventilated scenario remained thermally untenable in most of the second floor while the only room that remained untenable for victims in the PPV scenario is the fire room. Through the duration of the fire the addition of the PPV fan increased the amount of tenable area for victims and lowered temperatures in the attack path of the fire fighters.

Figure 5-24 displays comparisons of first floor temperatures 1.2 m (4 ft) from the floor in two hundred second intervals for 600 s. The first floor remained at ambient temperatures for the duration of the simulation for both ventilation scenarios.

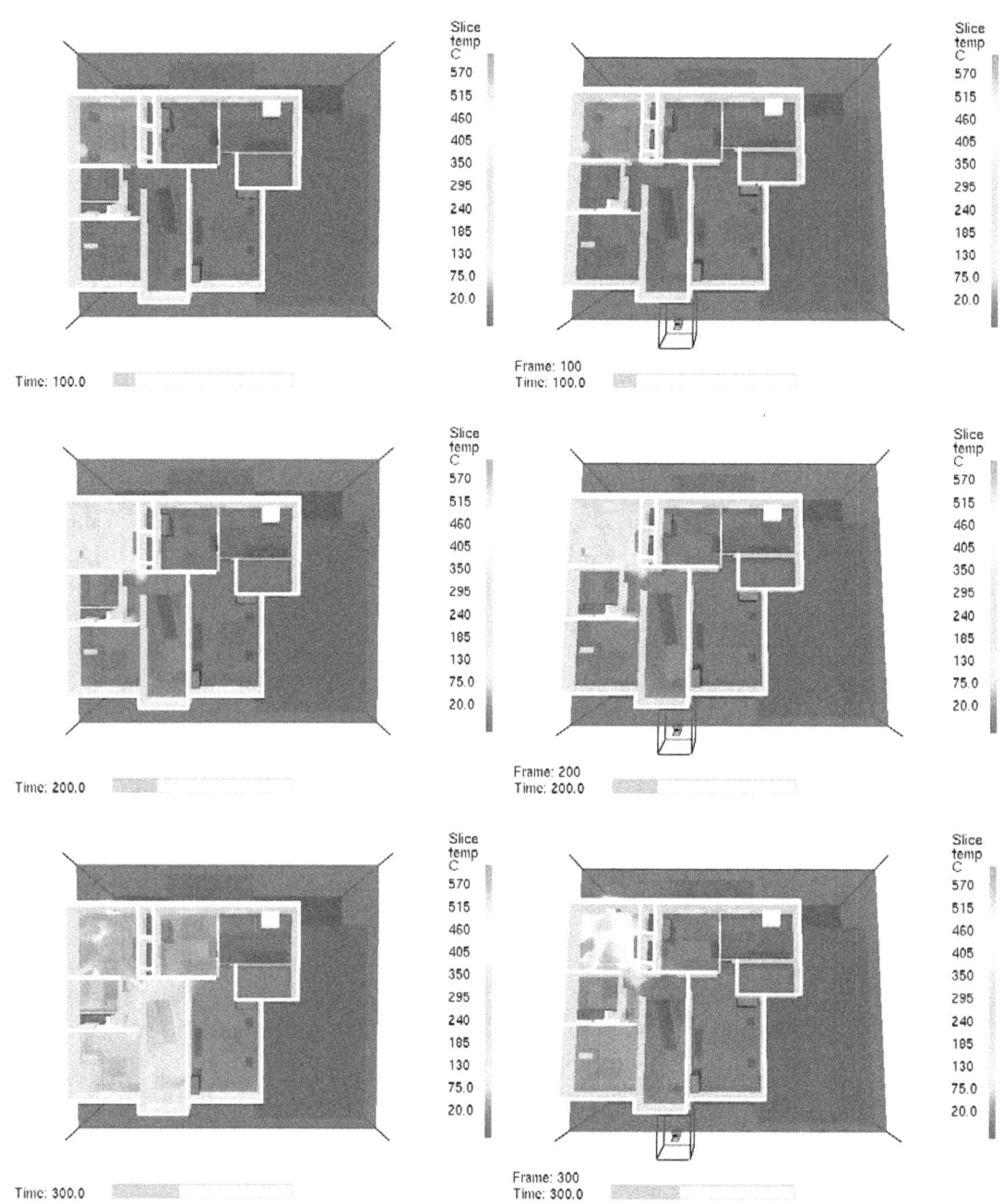

Figure 5-22. Comparison of Second Floor Temperatures 100 s to 300 s, Naturally Ventilated (Left) and PPV Ventilated (Right)

Figure 5-23. Comparison of Second Floor Temperatures 400 s to 600 s, Naturally Ventilated (Left) and PPV Ventilated (Right)

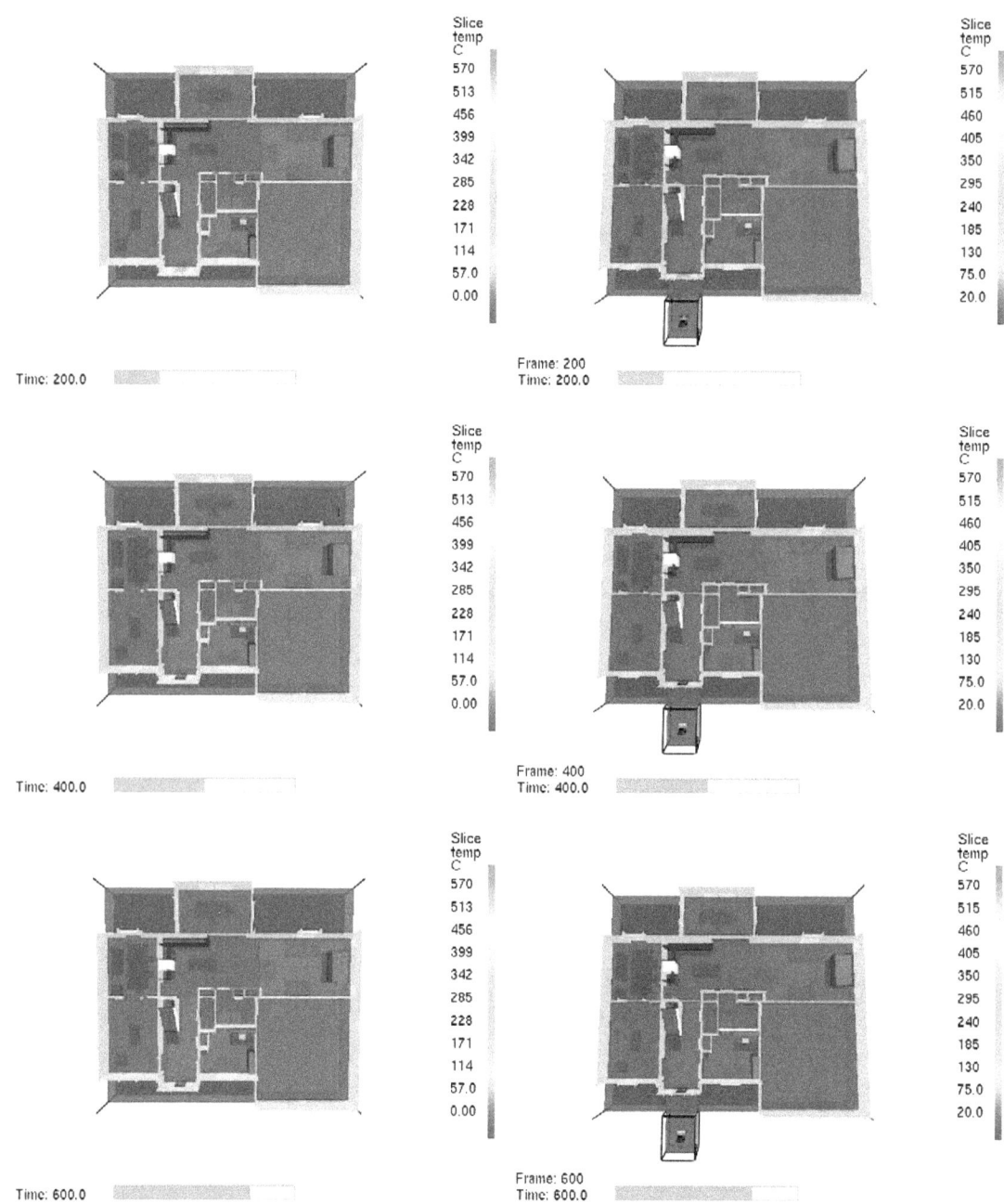

Figure 5-24. Comparison of First Floor Temperatures 200 s to 600 s, Naturally Ventilated (Left) and PPV Ventilated (Right)

5.3.4 Oxygen

Oxygen volume fraction concentrations were also examined in the simulations to assess the tenability conditions that existed in the house during the evolution of the fire. This horizontal slice was located at the same elevation as the temperature slice, 1.2 m (4 ft) above the floor. This analysis utilized a volume fraction of 12 % as the oxygen tenability threshold for victims [21]. It was assumed that fire fighters utilized self contained breathing apparatus. Therefore in this analysis, no oxygen tenability threshold was considered for fire fighters.

Figures 5-25 and 5-26 show the second floor oxygen volume fractions 1.2 m (4 ft) above the floor in 100 s intervals for 600 s. For the first 200 s the entire second floor was tenable for both ventilation scenarios. After ventilation at 300 s the fire was much larger and consuming more oxygen even though the window and front door were open. In the naturally ventilated scenario the left hand side of the second floor and the stairwell were approaching untenable conditions while the PPV ventilated scenario remained tenable everywhere but the fire room.

By 400 s the entire second floor in the naturally ventilated scenario was untenable with the front left bedroom as low as 5 % oxygen. This same room along with the fire room in the PPV ventilated scenario reached un-tenability, but the rest of the floor remained tenable. The front left bedroom oxygen concentration dropped so low because it is adjacent to the fire room and all its oxygen is consumed by the fire but there was no supply because the window was closed and the fire room drew all the oxygen from the front door.

The final two comparisons show the fire in the decay stage as the oxygen levels rise. The PPV ventilated scenario returned tenability to the second floor much more rapidly than the naturally ventilated scenario. This was expected as the fan was forcing fresh air into the house. This increased oxygen increases the survivability of victims and allowed fire fighters to perform overhaul procedures sooner.

Figure 5-27 displays comparisons of first floor oxygen volume fractions 1.2 m (4 ft) from the floor in 200 s intervals for 600 s. The first floor remained at ambient oxygen levels for the duration of the simulation for both ventilation scenarios.

Figure 5-25. Comparison of Second Floor Oxygen Volume Fractions 100 s to 300 s, Naturally Ventilated (Left) and PPV Ventilated (Right)

Figure 5-26. Comparison of Second Floor Oxygen Volume Fractions 300 s to 600 s, Naturally Ventilated (Left) and PPV Ventilated (Right)

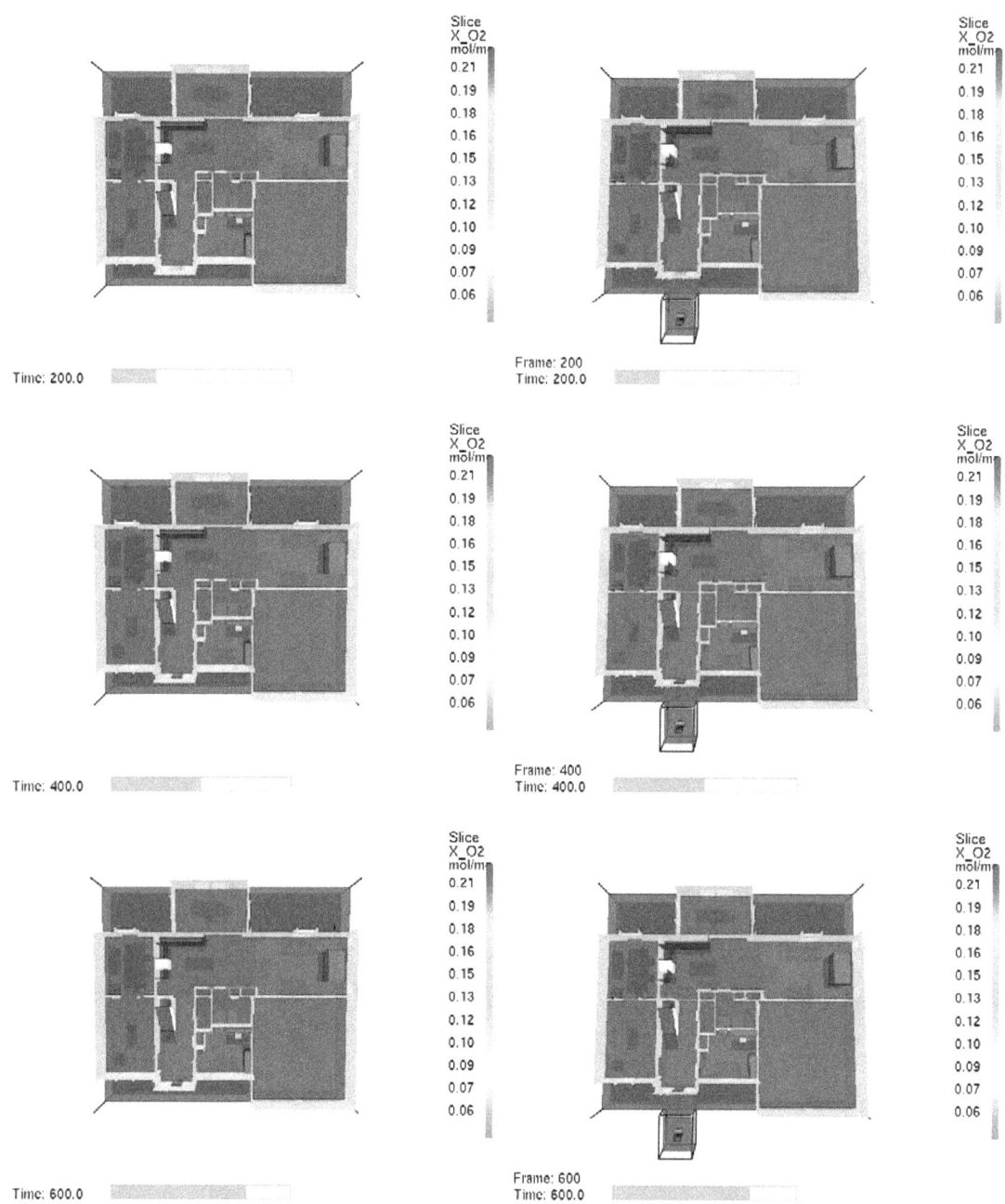

Figure 5-27. Comparison of First Floor Oxygen Volume Fractions 200 s to 600 s, Naturally Ventilated (Left) and PPV Ventilated (Right)

5.3.5 Velocity

Figures 5-28 through 5-34 are velocity slice files taken at 320 s, 80 s after the window and front door were opened. Figure 5-28 is located vertically through the center of the fan. This slice shows the cone of air described in chapter 1 and how it covers the entire height of the door. The velocity range was consistent with that achieved in chapter 2 (Figure 2-15).

Figure 5-29 and Figure 5-30 compare the velocities out of the fire room window. From this comparison it can be seen that the velocity was increased by the addition of the PPV fan. The fan increased the flow from approximately 4 m/s (13 ft/s) to 8 m/s (26 ft/s) which was consistent with the effect of the fan seen in chapter 4. Comparing figures 4-39 and 4-40 show how the experimental gas velocities were roughly doubled with the addition of the fan.

Figures 5-31 through 5-34 compare the velocities on the first and second floors 1.2 m (4 ft) above the floor at 320 s. The addition of the fan only slightly increased these mid-level velocities through the fire room; however, what is not shown is the direction of flow. Much of the flow in the naturally ventilated scenario was out of the door and into the hallway while the PPV ventilated flow was forced back into the fire room. On the first floor the difference was obvious with the addition of the PPV fan. It can also be observed that some of the flow created by the PPV fan circulated back to the front door. This could be due to the increased pressure and could be affected by the size of the exhaust vent.

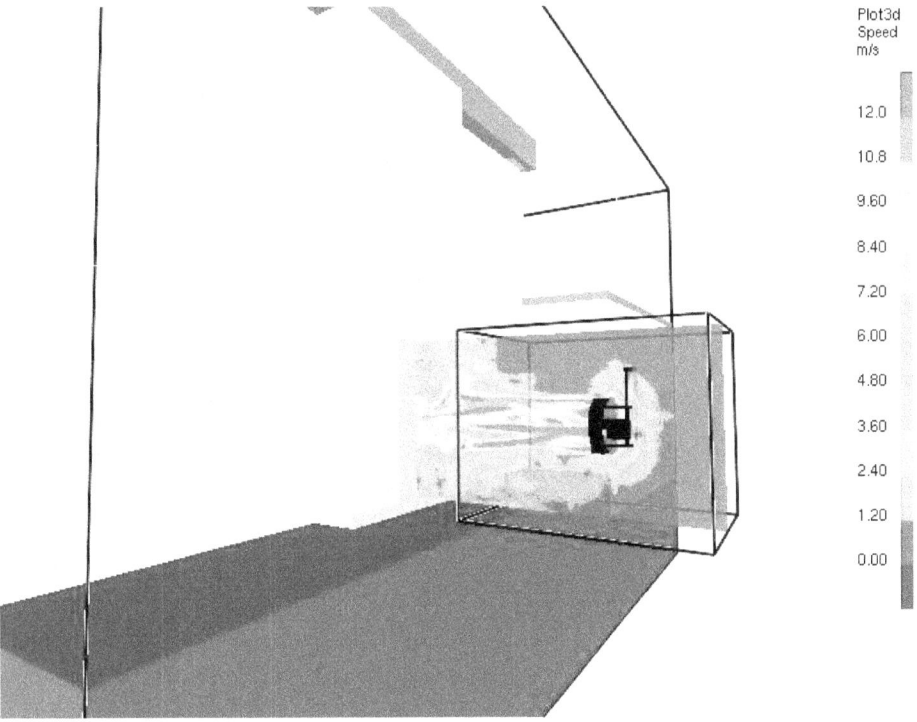

Figure 5-28. Vertical Velocity Slice Through Center of Fan

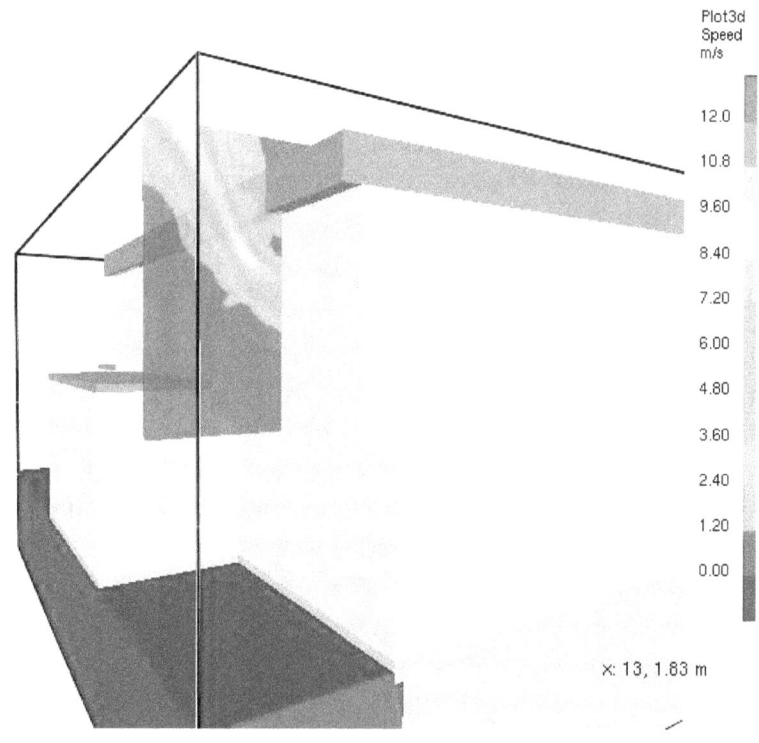

Figure 5-29. Velocity Slice Through Center of Vent Window During PPV Ventilation Scenario at 320 s

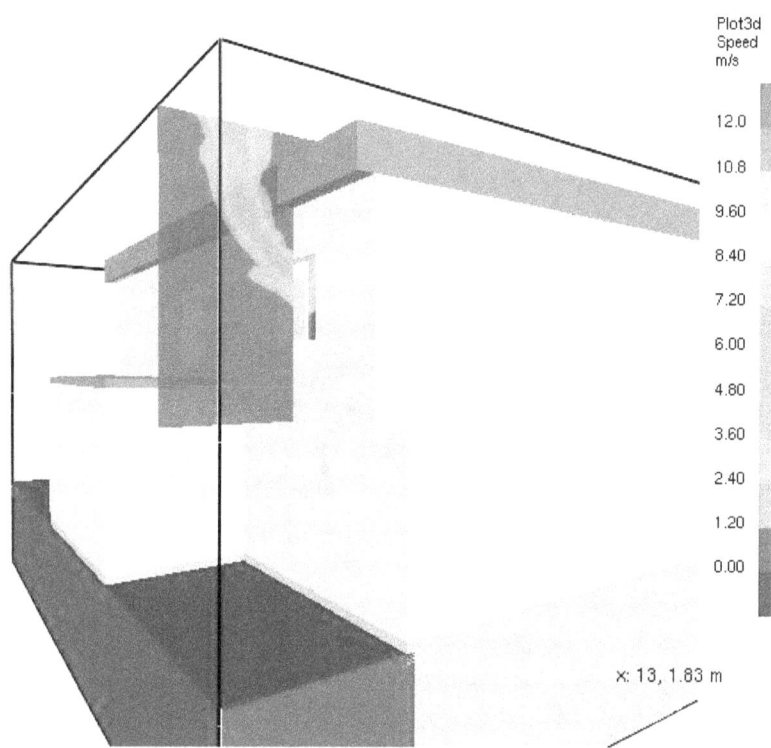

Figure 5-30. Velocity Slice Through Center of Vent Window During Natural Ventilation Scenario at 320 s

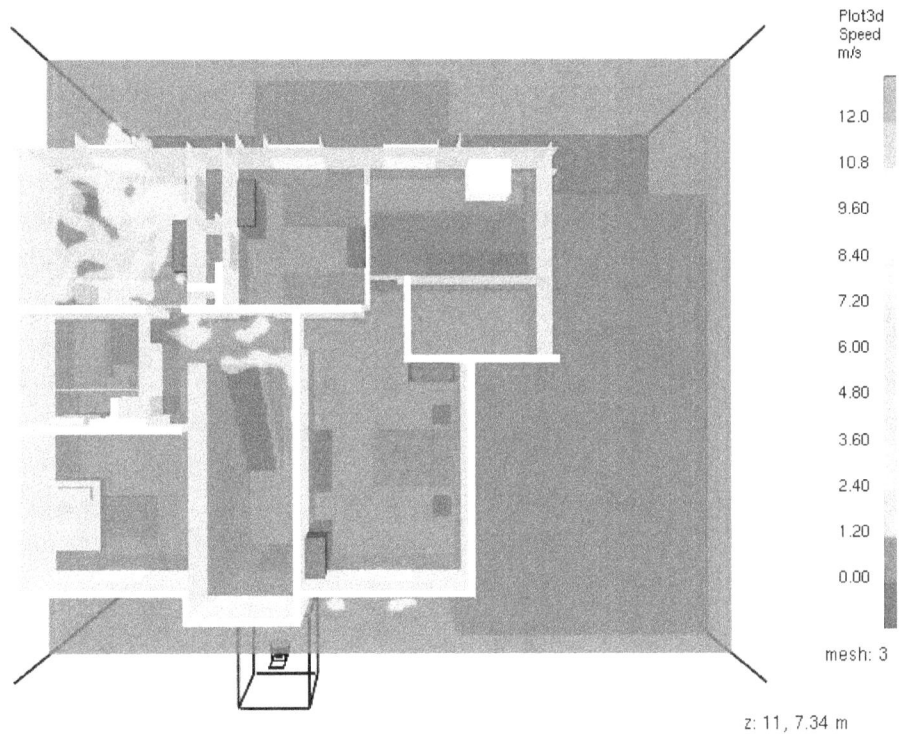

Figure 5-31. Horizontal Velocity Slice Through Second Floor During PPV Ventilation Scenario at 320 s

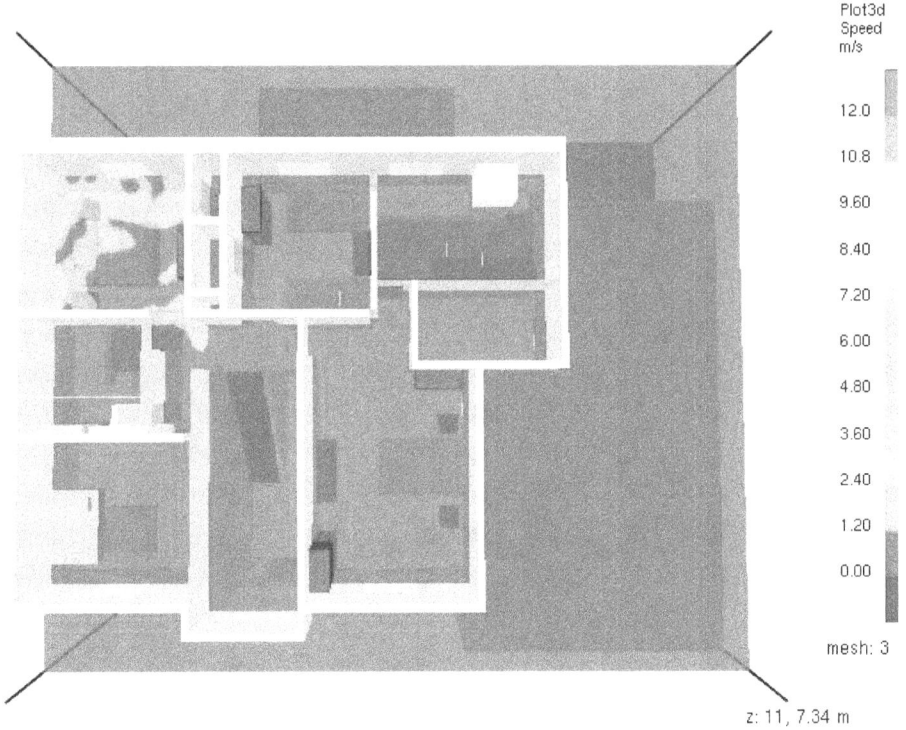

Figure 5-32. Horizontal Velocity Slice Through Second Floor During Natural Ventilation Scenario at 320 s

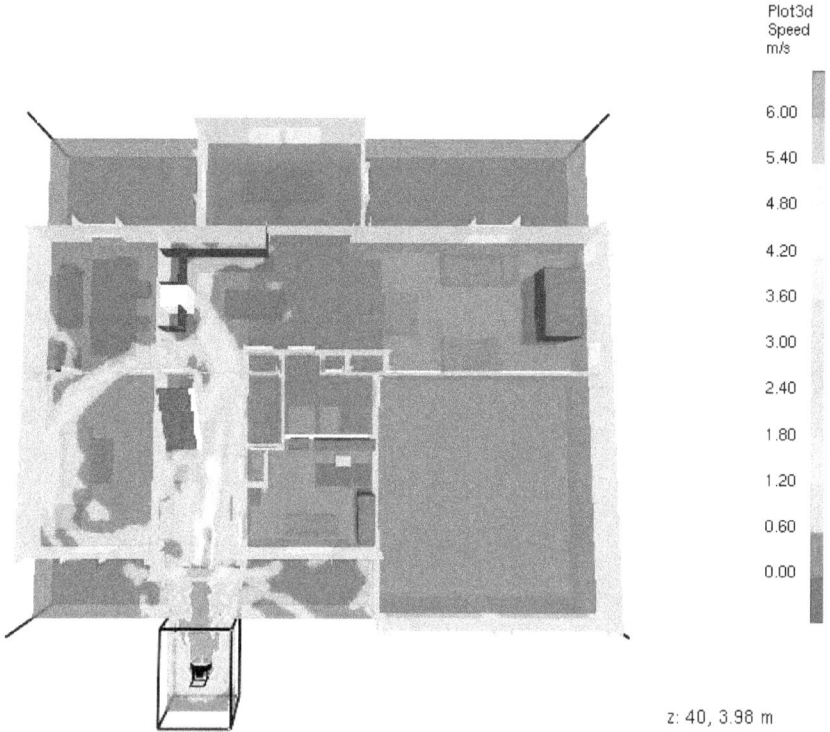

Figure 5-33. Horizontal Velocity Slice Through First Floor During PPV Ventilation Scenario at 320 s

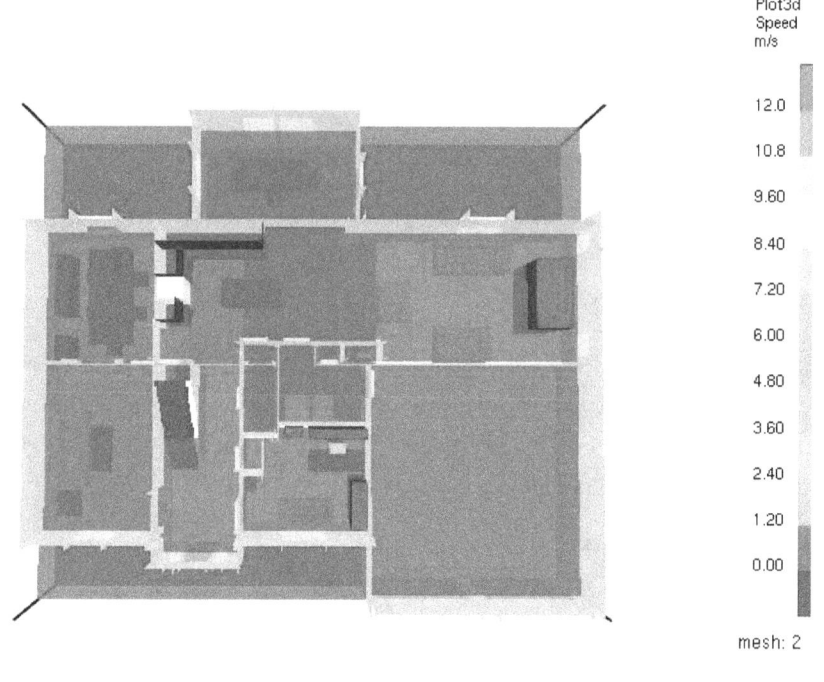

Figure 5-34. Horizontal Velocity Slice Through First Floor During Natural Ventilation Scenario at 320 s

5.4 Colonial House Summary

Two simulations were run for 800 s to examine the effects of a positive pressure ventilation fan on fire spread, smoke movement, temperature, oxygen concentration and velocities in a colonial house. These parameters were compared to smaller scale experimental data and published tenability criteria. There wais no direct experimental data that can be used for comparison; however the trends from the room fire experiments were a good guideline to understanding the accuracy of the simulations.

Prior to ventilation, fire growth and smoke spread were identical for both simulations. After ventilation the smoke movement out of the eaves and the flames from the fire room window were intensified by the positive pressure ventilation fan. Both of these observations were consistent with the room fire experiments in chapter 4.

The heat release rate from both simulations peaked at the same maximum value. This is not consistent with the room fire experiments but could be due to the longer path between the fan and the fire. After this peak heat release rate the PPV fan caused an increased heat release rate for approximately 70 s which is consistent with the room fire experiments. After this 70 s of increased burning the two curves converged just as was seen in the experiments.

In the PPV scenario gas temperatures were lower in all of the rooms on the second floor with the exception of the fire room than the temperatures in the naturally ventilated scenario. By 400 s most of the second floor was thermally untenable for victims in the naturally ventilated scenario while only one additional room next to the fire room reached the tenability limit in the positive pressure ventilated scenario. The addition of the PPV fan also lowered temperatures in the path to the fire room that would be used by fire fighters to attack the fire. The only room thermally untenable for fire fighters was the fire room. Temperatures on the first floor remained ambient for the duration of both simulations.

Oxygen levels followed similar trends as the temperature. As the temperature increased the oxygen levels typically decreased, as would be expected. Oxygen tenability limits were exceeded on the entire second floor during the naturally ventilated scenario by 400 s and was maintained until 600 s. Limits were only reached in the fire room and adjacent front bedroom during the PPV ventilated scenario. At 600 s the increased flow created by the PPV fan created tenable oxygen limits on the entire second floor. Oxygen levels remained ambient on the first floor for the duration of both simulations.

Velocity slices showed that the fan itself created flow at the doorway that was of the proper magnitude and direction to create the cone of air necessary for positive pressure ventilation. These values were also consistent with laboratory experiments in Chapter 2. Much of the flow in the naturally ventilated scenario is out of the fire room door and into the second floor hallway while the PPV ventilated flow forced the fire gases back into the fire room. On the first floor the difference was obvious with the addition of the PPV fan. It was also observed that some of the flow created by the PPV fan circulated back to the front door. This could be due to the increased pressure and could be affected by the size of the exhaust vent.

The results of this comparison show that the PPV fan created tenable conditions in the house that would not normally be there under natural ventilation tactics. This is only one ideal scenario with well coordinated ventilation tactics assumed. These simulations follow the fire dynamics seen in laboratory experiments and provide confidence that the positive pressure ventilation tactic can be analyzed without the need for countless full-scale field tests. If full-scale opportunities arise in the future they could be beneficial as there is no true replacement for such data. Countless ventilation scenarios can be performed on this colonial style house at the expense of computational time, examining variables such as ventilating other rooms, ventilating at different times and starting the fire in other rooms. As computers continue to advance in speed and capability this cost will decrease allowing for a more timely result of the impact of the PPV fan. Simulating more scenarios would allow for a better understanding of the tactic, better training and potentially could assist in equipment design.

Chapter 6: Uncertainty

6.1 Experimental

There are different components of uncertainty in the length measurements, gas temperatures, mass of fuel packages, gas velocity and heat release rate data reported here. Uncertainties are grouped into two categories according to the method used to estimate them. Type A uncertainties are those which are evaluated by statistical methods, and Type B are those which are evaluated by other means [22]. Type B analysis of systematic uncertainties involves estimating the upper (+ a) and lower (- a) limits for the quantity in question such that the probability that the value would be in the interval (± a) is essentially 100 %. After estimating uncertainties by either Type A or B analysis, the uncertainties are combined in quadrature to yield the combined standard uncertainty. Multiplying the combined standard uncertainty by a coverage factor of two results in the expanded uncertainty which corresponds to a 95 % confidence interval (2σ).

Components of uncertainty are tabulated in Table 10 (chapter 2 and 3 experiments) and Table 11 (chapter 4 experiments). Some of these components, such as the zero and calibration elements, are derived from instrument specifications. Other components, such as radiative cooling/heating, include past experience with thermocouples in high temperature fuel rich environments.

Each length measurement was taken carefully but due to some construction issues such as the size and straightness of the lumber, the slots for the strings on the grid having thickness, and the symmetry of the rather large room the total expanded uncertainty was estimated to be 6 %. The flow measurements were taken in the complex flow of the positive pressure ventilation fan, which created a total expanded uncertainty of 14 % primarily due to the repeatability and the randomness of the measurements.

The uncertainty in the upper layer gas temperature measurements includes radiative cooling in each of the experimental series, but also includes radiative heating for thermocouples located in the lower layer of the full-scale experiments. Pitts et al. [13] quantified the errors of bare bead thermocouples as ranging from 7 % in the hot upper gas layer to as much as 75 % in the lower layer. The potential for large errors in the lower layer are a function of the effective temperature of the surroundings. In cases where the effective temperature of the surroundings is high, the error can be more significant. In cases, similar to a developing fire in a compartment, the temperature measurement errors in the lower layer are smaller as the fire develops through flashover, since the effective temperature of the floor and walls are relatively cool. Post-flashover, the potential for measurement error increases as the temperature of the surroundings increase. Small diameter thermocouples were used to limit the impact of radiative heating and cooling. This resulted in an estimate of ±15 % total expanded uncertainty in temperature measurements.

Differential pressure reading uncertainty components are derived from pressure transducer instrument specifications. The transducers were factory calibrated and the zero and span of each was checked in the laboratory prior to the experiments. The readings from the pressure transducers were used to generate gas velocities.

Load cells were utilized to measure fuel package mass. The load cell was calibrated with a standard mass prior to recording the mass of each fuel item. After obtaining mass data on each of the fuel components, items were selected at random to be reweighed in order to estimate repeatability.

Total expanded uncertainties associated with oxygen calorimetry techniques are discussed in greater detail by Bryant et al. [12]. This uncertainty was estimated to be 11 % and included components derived from gas concentrations, temperature and gas flows. This estimation is based on the calorimetry system alone and does not account for the uncertainty that exists due to the experimental configuration. There is a delay time for the combustion gases to reach the hood and calorimetry instrumentation. The heat released within the fire room has an additional uncertainty associated with it. This uncertainty varies during the experiment. After the window was opened the uncertainty was minimized due to the majority of the burning occurring outside of the room.

Table 10. Mapping and Simple Room Experimental Uncertainty

Component	Standard Uncertainty	Combined Standard Uncertainty	Total Expanded Uncertainty

Length Measurements
    Grid Size
    String Location
    Fan Stand Height
    Anemometer Location
    Fan Location
    Room Dimensions
    Repeatability
    Random
	
	± 1 %
	± 0.5 %
	± 0.5 %
	± 1 %
	± 1 %
	± 1 %
	± 2 %
	± 2 %
		3 %	6 %

Flow Measurements
    Calibration
    Repeatability
    Random
	
	± 0.5 %
	± 5 %
	± 5 %
		7 %	14 %

Note: Random and repeatability evaluated as Type A, other components as Type B			

Table 11. Room Fire Experimental Uncertainty

	Component Standard Uncertainty	Combined Standard Uncertainty	Total Expanded Uncertainty

Gas Temperature
    Calibration[23]
    Radiative Cooling
    Radiative Heating
    Repeatability[1]
    Random[1]
	± 1 %
	- 5 % to + 0 %
	- 0 % to + 5 %
	± 5 %
	± 3 %
		- 8 % to + 8 %	- 15 % to + 15 %

Differential Pressure
    Calibration[11]
    Accuracy[11]
    Repeatability[1]
    Random[1]
	± 2 %
	± 1 %
	± 5 %
	± 5 %
		- 8 % to + 8 %	- 15 % to + 15 %

Mass of Fuel Package
    Zero
    Calibration
    Repeatability[1]
    Random[1]
	± 0.02 %
	± 1 %
	± 5 %
	± 3 %
		± 6 %	± 12 %

Length Measurements
    Instrumentation
        Locations
    Furniture
    Dimensions
    Fan Location
    Room Dimensions
    Repeatability[1]
    Random[1]
	± 1 %
	± 1 %
	± 1 %
	± 1 %
	± 2 %
	± 2 %
		± 3 %	± 6 %

Notes: 1. Random and repeatability evaluated as Type A, other components as Type B.			

6.2 Fire Dynamics Simulator

FDS can provide valuable insight into how a fire may develop or how the combustion gases will move throughout a structure. However the model is only a simulation. The model output is dependent on a variety of input and default values such as computational cell size, material properties, geometry, and vents.

The ability of the FDS model to predict accurately the temperature and velocity of fire gases has been previously evaluated by conducting experiments, both lab-scale and full-scale, and measuring quantities of interest. For relatively simple fire driven flows, such as buoyant plumes and flows through doorways, FDS predictions are within the experimental uncertainty of the values measured in the experiments [21]. For example, if a gas flow velocity is measured at 0.5 m/s (2 ft/s) with an experimental uncertainty of $\pm$ 0.05 m/s ($\pm$ 0.2 ft/s), the FDS model gas flow velocity predictions were also within the range between 0.45 m/s (1.5 ft/s) and 0.55 m/s (1.8 ft/s).

In large-scale fire tests reported in [24], FDS temperature predictions were found to be within 15% of the measured temperatures and the FDS heat release rates were predicted to within 20% of the measured values. Therefore the results are presented as ranges to address these uncertainties.

These experiments indicate that using the correct set of inputs, FDS is able to model the flows of positive pressure ventilation fans within a useful level of uncertainty.

Chapter 7: Conclusion

Data from three sets of full-scale experiments were compared with simulations completed with the computational fluid dynamic model Fire dynamics simulator (FDS). The full-scale experiments characterized a Positive Pressure Ventilation (PPV) fan in an open atmosphere, in a simple room geometry and during a room fire. Experimental results were visually and numerically compared to FDS results. The comparison showed reasonable agreement. A concluding scenario was modeled utilizing the calibration of the full-scale experiments to examine the effects of PPV on a fire in a two-story colonial-style house.

Numerous geometries were experimented within FDS to obtain a fan that provided both numerical and visual comparisons that were accurate. The positive pressure ventilation fan mapped in the open atmosphere yielded an average velocity difference between the experiment and the simulation of less than 10 %. Visually the experimental flow obtained by using the smoke generator compared well to the FDS simulation. Special attention must be given to the grid cell size, grid location and fan design to accurately model the fan flow.

The simple room geometry experiments added a change in flow direction and a pressure gradient to challenge the computer model. Experimental and simulation comparisons gave a velocity difference of 16.5 % at the window. The air movement at the window is the result of the pressure change in the room, interaction of the geometry of the room and the airflow. Qualitatively the flow visualized using the smoke generator matched the simulation very well.

The room fire experiments incorporated the interaction between the fan flow and a fire. The naturally ventilated and the positive pressure ventilated comparisons showed the fan created higher fire room temperatures, increased window gas flows and higher pressure differentials. The experiments also showed that the PPV fan created a 60 % increase in burning rate during the potential time of fire department attack. The limited set of data also provided insight to fire fighting tactics including the necessity of coordination of fire fighting crews to carry out positive pressure ventilation in the attack stages of a fire and the recommendation to delay advancement towards the fire until conditions created by the fan stabilized 60 s to 120 s after ventilation.

Computationally, visual and numerical comparisons demonstrate that the fire behavior of a room fire both with and without positive pressure ventilation can be modeled by FDS and visualized with Smokeview. Differences exist in the geometry and material properties but the fire dynamics and the net impact of the positive pressure ventilation fan can be captured with a reasonable degree of accuracy.
In order to expand the understanding of the effects of positive pressure ventilation it would be very informative to perform experiments in various types of full-scale structures. These buildings of opportunity are very difficult to obtain and very expensive to instrument. Using the lab scale tests documented above as a calibration

for the Fire dynamics simulator it was possible to visualize the effects of positive pressure ventilation fan on a colonial style house with a fire on the second floor.

The results of this comparison show that the PPV fan created tenable conditions in the house that would not normally be there under natural ventilation tactics. This is only one ideal scenario with well coordinated ventilation tactics assumed. These simulations follow the fire dynamics seen in laboratory experiments and provide confidence that the positive pressure ventilation tactic can be analyzed with reduced need for full-scale field tests. If full-scale opportunities arise in the future they would be beneficial as there is no true replacement for such data. A wide range of ventilation scenarios can be performed on this colonial style house at the expense of computational time. As computers continue to advance in speed and capability this cost will decrease allowing for a more timely result of the impact of the PPV fan.

The ability to model the effects of PPV given the proper input parameters in FDS provides a technique to add to the current deficient but fast growing understanding of positive pressure ventilation. Future research should incorporate FDS to answer specific fire department concerns with the use of PPV so that this tool can be utilized safely and effectively.

References

1. Carlson, G., "Volunteer's Corner, Positive Pressure Ventilation, Some Questions," *Fire Engineering*, March 1989, p. 9.

2. Coleman, J., "Roundtable, Positive Pressure Ventilation," *Fire Engineering*, August 1999, p. 22-32.

3. Yates, M., "Positive Pressure Ventilation - The Wind of Change," *BCC*. United Kingdom. 2001.

4. Tempest Technology Corporation, http://www.tempest-edge.com, January 2003.

5. McGrattan, K. B., Baum, H. R., Rehm, Ronald G., Hamins, Anthony, Forney, Glenn P., "Fire Dynamics Simulator – Technical Reference Guide," National Institute of Standards and Technology, Gaithersburg, MD., NIST Special Publication 1018, December 2004.

6. McGrattan, K. B., Forney, G. P., "Fire Dynamics Simulator – User's Manual," National Institute of Standards and Technology, Gaithersburg, MD., NIST Special Publication 1019, December 2004.

7. Forney, G. P., McGrattan, K. B. "User's Guide for Smokeview Version 4 – A Tool for Visualizing Fire Dynamics Simulation Data," National Institute of Standards and Technology, Gaithersburg, MD., NIST Special Publication 1017, August 2004.

8. Omega Engineering Inc., "User's Guide for the HH-31A Handheld Anemometer," Stamford, CT, 1996.

9. Hall, R, and Adams, B., *Essentials of Fire Fighting, 4th ed.*, Stillwater: Oklahoma State University, 1998.

10. Super Vacuum Manufacturing Company, Inc., http://www.supervac.com, January 2003.

11. Setra Systems, Inc., "Installation Guide, Setra Systems Model 264 Differential Pressure Transducer," Boxborough, MA., 1999.

12. Bryant, R.A., Ohlemiller, T.J., Johnsson, E.L., Hamins, A.H., Grove, B.S., Guthrie, W.F., Maranghides, A., and Mulholland, G.W., "The NIST 3 MW Quantitative Heat Release Rate Facility", NIST Special Publication 1007, National Institute of Standards and Technology, Gaithersburg, MD, 2003.

13. Pitts, W.M., E. Braun, R.D. Peacock, H.E. Milter, E.L. Johnsson, P.A. Reneke, and L.G. Blevins, "Temperature Uncertainties for Bare-Bead and Aspirated

Thermocouple Measurements in Fire Environments," *Thermal Measurements: The Foundation of Fire Standards*. American Society for Experimenting and materials (ASTM), Proceedings. ASTM STP 1427, December 3, 2001, Dallas, TX.

14. Blevins, L.G. "Behavior of Bare-bead and Aspirated Thermocouple in Compartment Fires." Proceedings of the 33rd National Heat Transfer Conference August 15-17, 1999, Albequerque, New Mexico.

15. Stott, R. "Report on PPV Trials at Oxford Road, Preston, U.K." p. 10-14 January, 2000.

16. Svensson, S. "Experimental Study of Fire Ventilation During Fire Fighting Operations", *Fire Technology*, Vol. 37, 2001, p. 69-85.

17. Ezekoye, O., Lan, C., Nicks, R. "Positive Pressure Attack for Heat Transport in a House Fire," The 6th ASME-JSME Thermal Engineering Joint Conference. March 16-20, 2003.

18. Gojkovic and Bengtsson. "Some Theoretical and Practical Aspects on Fire Fighting Tactics in a Backdraft Situation," http://www.firetactics.com/DANIEL-GOJKOVIC.htm October 2005.

19. Krasny, J., Rockett, J. A., and Huang, D. "Protecting Fire Fighters Exposed in Room Fires: Comparison of Results of Bench Scale Test for Thermal Protection and Conditions During Room Flashover." *Fire Technology, p.* 5-19, (1988).

20. *Protective Clothing for Structural Firefighting,* NFPA Standard 1971, National Fire Protection Association, Quincy, MA (2003).

21. McGrattan, K. B., Hamins, A., Stroup, D., "Sprinkler, Smoke & Heat Vent, Draft Curtain Interaction – Large Scale Experiments and Model Development," National Institute of Standards and Technology, Gaithersburg, MD., NISTIR 6196-1, September 1998.

22. Taylor, B. N. and Kuyatt, C. E., "Guidelines For Evaluating and Expressing the Uncertainty of NIST Measurement Results." National Institute of Standards and Technology (U.S.) NIST-TN 1297; pp. 20 September, 1994.

23. Omega Engineering Inc., The Temperature Handbook, Vol. MM, pages Z-39-40, Stamford, CT., 2004.

24. McGrattan, K. B., Baum, H. R., Rehm, R.G., "Large Eddy Simulations of Smoke Movement," *Fire Safety Journal*, vol 30 (1998), p. 161-178.

25. Babrauskas, V. Ignition Handbook. Fire Science Publishers. Issaquah, WA 2003.

26. Babrauskas, V. Burning Rates. *The SFPE Handbook of Fire Protection Engineering,* National Fire Protection Association, Quincy, MA, Third Edition (2002).

27. Krasny, J. F., Rockett, J. A., Huang, D. "Protecting Fire Fighters Exposed in Room Fires. Part 1. Comparison of Results of Bench Scale Test for Thermal Protection and Conditions During Room Flashover." Clemson University. Protective Clothing--An Update and Overview of Personal Protection Against Chemical, Thermal and Nuclear Hazards. 1st Annual Conference. May 27-28, 1987, Clemson, NC, 1-28 pp, 1988.Fire Technology, Vol. 24, No. 1, 5-19, February 1988.

www.ingramcontent.com/pod-product-compliance
Lightning Source LLC
Chambersburg PA
CBHW081725170526
45167CB00009B/3700